The Quality of the Archaeological Record

The Quality of the Archaeological Record

CHARLES PERREAULT

The University of Chicago Press
Chicago and London

The University of Chicago Press, Chicago 60637
The University of Chicago Press, Ltd., London
© 2019 by The University of Chicago
Published 2019

28 27 26 25 24 23 22 21 20 19   1 2 3 4 5

ISBN-13: 978-0-226-63082-3 (cloth)
ISBN-13: 978-0-226-63096-0 (paper)
ISBN-13: 978-0-226-63101-1 (e-book)
DOI: https://doi.org/10.7208/chicago/9780226631011.001.0001

Library of Congress Cataloging-in-Publication Data

Names: Perreault, Charles, 1980– author.
Title: The quality of the archaeological record / Charles Perreault.
Description: Chicago ; London : The University of Chicago Press, 2019. |
    Includes bibliographical references and index.
Identifiers: LCCN 2018045365 | ISBN 9780226630823 (cloth : alk. paper) |
    ISBN 9780226630960 (pbk. : alk. paper) | ISBN 9780226631011 (e-book)
Subjects: LCSH: Archaeology—Data processing. | Archaeology—Methodology. |
    Antiquities—Collection and preservation.
Classification: LCC CC80.4 .P47 2019 | DDC 930.1028/2—dc23
LC record available at https://lccn.loc.gov/2018045365

Contents

Preface

I can trace the inception of this book to a very specific moment and place: a Friday afternoon in June 2009, UCLA campus, in Los Angeles, California. Back then, I was a graduate student and was meeting with Robert Boyd, a faculty member in my department.

At the time, I was enthralled by cultural evolution theory (not the old sociocultural kind but the dual-inheritance sort). I was ready to run with it, all gas, no brakes, and apply it to the archaeological record. Rob is one of the early architects of cultural evolution theory, so he was naturally added to my committee and put in charge of the theory part of my qualifying exams. Early in the meeting, he told me that for my exam I would have to discuss whether archaeological data can be used to detect the routes of cultural transmission, transmission biases, or the importance of social learning relative to other modes of learning (all things I wanted to study archaeologically). I had the weekend to write an essay and answer his question.

By Saturday, my answer to Rob's question had morphed from an "of course it can" to a humbler "actually maybe not." And by Sunday evening I had lost faith in much of what I thought archaeology was about. After a weekend of thinking hard about what it means to answer a question scientifically and reading dozens of articles from paleontologists struggling to reconcile the fossil record with evolutionary genetics, I had come to see how large the gulf is that separates the archaeological record and the microevolutionary processes described by cultural evolution theory. Too large, I thought, to be ever bridged, at least in a way that I would find valid and reasonable. Yet, I didn't despair. Quite the contrary: I was thrilled. The same paleontologists who had long stopped slavishly interpreting their data in microevolutionary terms had been doing all sorts of exciting things with the fossil record, such as studying patterns in rates of evolutionary change and trends in taxonomic diversity of extinction rates—topics that were not only fascinating but also well suited to the quality of the fossil record. The same kind of approach could also be adopted by archaeologists. I felt like I had hit upon an untapped vein of gold that, I suspected, ran deep and wide under the ground.

In the end, I would write a dissertation on a different topic. But the question of the quality of the archaeological record remained at the back of my mind, and I returned to it immediately after moving to the Santa Fe Institute, where I had been offered an Omidyar Postdoctoral Fellowship. I realized very quickly that the critique I laid out in my exam essay extended well beyond the domain of cultural evolution theory. And I would also realize soon enough that others before me had ventured into the same territory, chief among them Geoff Bailey with his "time perspectivism" approach. He, and many others who have followed in his footsteps—Stein, Murray, Wandsnider, Holdaway, Shott, to name just a few—have deeply shaped my thoughts as I was writing this book. Theirs are the shoulders upon which I stand.

This book also owes a large debt to Jeff Brantingham. Jeff taught me that the archaeologist's job is not only to study the content of the archaeological record but also to study the archaeological record *itself*. Jeff is also one of the most original thinkers I know. Not only does he think outside the proverbial box, but he turns it upside down and will not hesitate to throw it away if need be. The heavy dose of taphonomic and critical thinking that he bestowed on me lays the groundwork for everything that appears in this book. And his constant encouragements have kept me going when I was in a rut.

This book was completed over the course of several years, and I have benefited from dozens of conversations with various people. Perhaps unbeknownst to them, and though they may not agree with the content of this book, in whole or in part, the following people have inspired me, pointed me in new directions, helped me spot some of the weaker links in my arguments, or forced me to think and write more clearly. I thank them all: Michael Barton, Deanna Dytchkowskyj, Doug Erwin, Marcus Hamilton, Erella Hovers, Tim Kohler, Steve Kuhn, Lee Lyman, David Madsen, Curtis Marean, David Meltzer, Kostalena Michelaki, Chris Morehart, Tom Morgan, Michael O'Brien, Scott Ortman, Jonathan Paige, Karthik Panchanathan, Matt Peeples, Luke Premo, Hannah Reiss, David Rhode, Eric Rupley, Jerry Sabloff, Michael Smith, Chip Stanish, Nicolas Stern, LuAnn Wandsnider, Meg Wilder, and two anonymous reviewers. A special thanks to Michael Shott, who generously reviewed the last two versions of the manuscript and provided me with thorough, challenging, but also constructive and supportive feedback. I also want to thank the editors at the University of Chicago Press: first, Christie Henry, who shepherded the book though the first phases of review, and then, Scott Gast, who saw it through the finish line. I am also indebted to Pamela Bruton, who copyedited the book and made it better in so many ways. Finally, this book was written while in residence in various institutions and benefited from their support: the Santa Fe Institute, the University of Missouri, and, my most recent home, Arizona State University.

The Search for Smoking Guns

Archaeologist Geoff Bailey (1981, 104) incisively observed that "archaeology . . . is reduced to an appendix, at best entertaining, at worst dispensable, of ecology, sociology, or whichever study of contemporary behaviour happens to be in current fashion." Although harsh, his comment is still accurate more than 35 years later. Bailey was referring to the problem of interpreting what he called macrotemporal trends (i.e., the archaeological record) in terms of microtemporal processes (such as those described by anthropological theory). Given how rarely archaeological research is cited by scientists outside archaeology, let alone outside anthropology, and given its low status within the academy (Upham, 2004), it does seem like the contribution of archaeology to our understanding of human behavior has been, for the most part, unimportant.

Archaeology has remained an appendix to the other sciences of human behavior because archaeologists have been insisting on interpreting archaeological remains in terms of microscale processes. For various historical, psychological, and training reasons, archaeologists have come to view themselves as prehistoric ethnographers, whose goal is to interpret the archaeological record in terms of processes borrowed from other disciplines, such as cultural anthropology, psychology, and economics. In doing so, they have been producing a flow of information about the human past that is impressive—and yet unverifiable and likely erroneous.

The processes borrowed by archaeologists operate over very short time scales—so much so that most of them are in fact irremediably *underdetermined* by the archaeological record. Underdetermination is related to the more familiar concept of "equifinality." Equifinality is a quality of processes: processes are equifinal when they lead to the same outcome and are observationally equivalent. Underdetermination, on the other hand, is a quality of our observations: a set of observations underdetermines a set of processes when it cannot discriminate between them. (The equifinality/underdetermination problem discussed in this book concerns what philosophers refer to as *local* underdetermination, which is the type of underdetermination that arises during the normal course of scientific practice. It does not refer to *global* underdetermination, which challenges the possibility of scientific knowledge by postulating that for every theory, a large, and possibly infinite, number of rivals that are empirically equivalent always exist. See Fraassen, 1980; Kukla, 1998; Turner, 2007.)

The term "equifinality" is typically reserved for processes that lead to the *exact same* outcome, such that it will never be possible to distinguish them statistically (von Bertalanffy, 1940, 1949), or for processes that are *difficult* to distinguish, either because we lack the observational or statistical tools to do so (Laudan and Leplin, 1991; A. Rogers, 2000) or because we have failed to define our research questions in concrete and operational terms (Binford, 2001). In contrast, the term "underdetermination" tends to be used to describe the situations in which two processes are equifinal not necessarily because they are impossible or difficult to distinguish but because the data at hand cannot distinguish between the processes that generated them. The underdetermination problem of archaeology comes from a discrepancy between the coarseness of archaeological data and the microscale nature of archaeological theories (Bailey, 1981). The larger this discrepancy is, the more archaeological data will underdetermine the various economic, psychological, and social processes that archaeologists purportedly study.

The very way archaeologists test hypotheses undermines their capacity to make valid inferences about the human past. Because of the underdetermination problem, archaeologists have not been successful at inferring past causes. Indeed, how many questions about the human past have archaeologists answered in a definitive manner? With the exception of plain-vanilla cultural historical questions, the answer is, very few. Why is that? Because archaeologists often settle on an explanation on the sole basis that it can be made *consistent* with their data, thereby ignoring the fact that there are a number of alternative explanations that are just as consistent with the data. The use of consistency as a criterion to test hypotheses has made archaeologists overconfident about what can be learned from the archaeological record and allowed them to turn a blind eye to the harsh reality that the archaeological record underdetermines most of its causes.

Ultimately, the capacity of archaeologists to infer past causes depends on the quality of the archaeological record—on how much information about past events has been preserved in nature. Unlike experimental scientists, historical scientists such as archaeologists cannot use laboratory methods to manufacture new empirical evidence or to shield themselves from false-positive or false-negative results. This strict dependence on the quality of the archaeological record is anything but trivial. It means that the archaeological record—not archaeologists—dictates what can and cannot be learned about the past. Over the next few chapters, I will show that archaeology's current research agenda overestimates the quality of the archaeological record and, facilitated by the way archaeologists have been testing hypotheses, has led them to a place where most of their research questions either remain forever unresolved or are settled with wrong answers. The only way out of this situation is to recalibrate the research program of the discipline so that it is commensurate with the quality of the archaeological record.

By recalibrating their research program to the quality of the archaeological record, archaeologists can not only produce epistemologically valid knowledge about the past but also discover genuinely novel and possibly theory-challenging processes. For instance, archaeologists can mine the global archaeological record to detect macroscale processes—processes that operate above the hierarchical level of the individual and at such a slow rate that their effect can be detected only from an observation window that is thousands of years long and thousands of kilometers wide. Discovering such macroscale processes, which are effectively invisible to other social scientists, would be a significant achievement and a major contribution of archaeology to our understanding of human behavior.

Experimental Sciences and Historical Sciences

Epistemological discussions about archaeology tend to emphasize the distinction between "history" and "science," the idea being that history and science constitute different intellectual paradigms that require different methods. Today, archaeology largely defines itself as a science, and oftentimes in opposition to history. Archaeology students are taught that processual archaeology, by shifting archaeologists' focus from historical particularisms to cross-cultural regularities, sought to elevate our discipline from the rank of mere history to the high pedestal of science.

But history and science are not alternative intellectual paths. Cultural historians and processual archaeologists are engaged in the same activity: explaining contemporary observations of the archaeological record (i.e., observations made in the present time) in terms of their past causes. This makes archaeology, of every theoretical flavor, fall squarely under the umbrella of historical sciences. To better understand how we can gain knowledge about the past, we need to appreciate how *historical sciences* work. This is best done by contrasting historical sciences with *experimental sciences*.

Experimental scientists can directly observe their phenomenon of interest and test hypotheses in the controlled environment of the laboratory. By manipulating the conditions of their experiments, they can bring about the test conditions specified by their hypotheses. They can also repeat their experiments to ensure consistent results. An even more important feature of their practice is that by controlling for extraneous factors in their experiments, they can shield their hypotheses from false-positive and false-negative results (Cleland, 2001; Jeffares, 2008). Thus, with the help of laboratory methods, experimental scientists can identify causal relationships by observing how different initial conditions generate different results—in other words, they go from causes to effects.

Historical scientists exploit the opposite direction of the causality chain: they go from effects back to causes, by explaining contemporary observations in terms of their past causes. The range of research endeavors encompassed by historical sciences is large and varies in scope from the very vast (how did our galaxy form?) to the minute (why did the space shuttle *Challenger* explode?) (Forber and Griffith, 2001). Archaeologists, astrophysicists, geologists, paleontologists, but also NASA engineers and detectives tasked to solve crimes, are all historical scientists.

Mirroring the distinction many archaeologists make between archaeology-as-science and archaeology-as-history, experimental and historical sciences are often contrasted in terms of their objects of study. Whereas experimental scientists tend to be interested in classes of objects (e.g., how do helium molecules, neurons, or viruses behave?), historical scientists are more likely to investigate token objects (e.g., *this* star, *this* volcano, *this* war) (Cleland, 2001; Tucker, 2011). There is some truth to this characterization, but in reality, both types of sciences interface with classes of objects and token objects (Turner, 2007), constantly going back and forth between particular historical cases and "ahistorical" generalizations (Eldredge, 1989; Trigger, 1978). Thus, historical sciences are defined, not by their object of study, but by the fact that their object of study is in the past.

Unlike experimental scientists, historical scientists cannot directly observe the phenomena that interest them as they unfold but can observe only their outcomes. They cannot replicate the past in a laboratory setting, either for practical reasons (the formation of a galaxy, the development of agriculture) or ethical reasons (mass extinction, epidemics), let alone manipulate it. Historical scientists do have laboratories and laboratory methods, but they serve a different purpose than in experimental sciences. Whereas experimental scientists use the laboratory to manufacture new empirical evidence and to bring about various test conditions, historical scientists use laboratories to expand their search for smoking guns. For instance, the archaeology laboratory is where field data are processed, cleaned, cataloged, and analyzed. More critically, they use laboratory apparatuses to expand the range of data they observe beyond the range of traces that can be observed in the field, like a count of pollen in a soil sample or the $^{14}C/^{12}C$ ratio in a bone fragment. Yet, archaeologists still lack recourse to *experimental methods*. In lieu of the experimentalists' clean, uncontaminated, and controlled laboratories, they are stuck, like other historical scientists, with whatever traces have been left by nature's messy experiments (Jeffares, 2008). More importantly, they cannot do the very thing that makes experimental sciences so powerful: control experimentally for factors that are extraneous to their hypothesis and that may lead to false-positive or false-negative results. Instead, they must resort to finding smoking guns hidden in nature.

How Historical Sciences Work: The "Smoking-Gun" Approach

Historical scientists have had their fair share of triumphs: the discovery of tectonic-plate drift, the reconstruction of Pleistocene climate, and the calculation of the age of the universe are amazing feats of scientific ingenuity. Somehow, historical science can work.

Historical scientists successfully learn about the past by employing a "smoking-gun" approach. They start by formulating multiple, mutually exclusive hypotheses and then search for a "smoking gun" that discriminates between these hypotheses (e.g., Cleland, 2001, 2002, 2011; Forber and Griffith, 2001; Jeffares, 2008, 2010; Tucker, 2011; Turner, 2005, 2007). A smoking gun is a piece of evidence, discovered through fieldwork, that discriminates unambiguously between the competing hypotheses. The smoking gun can be anything—it can be a singular trace like a radiocarbon date, a set of traces such as a ceramic assemblage, or something more abstract, like a statistical signal.

The smoking-gun approach to historical science is a three-stage process (Cleland, 2011) (fig. 1.1). First, a set of competing hypotheses to explain the traces found in the field is generated. Then, researchers conduct fieldwork in order to find a smoking gun. Third, when a smoking gun is found, the set of competing hypotheses is first culled and then augmented in the light of new evidence and advances in theory. And the search for a smoking gun starts again.

The study of the extinction of the dinosaurs provides an example of these three stages (Cleland, 2001). All nonavian dinosaurs went extinct about 65.5 million years ago (Alroy, 2008; Macleod et al., 1997). Before the 1980s, several explanations had been suggested to account for their demise, including a

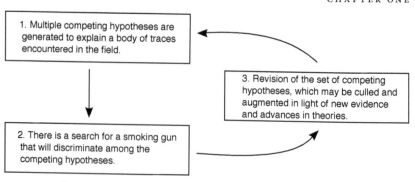

FIGURE 1.1: The three stages of prototypical historical research (Cleland, 2011).

meteorite impact, climate change, magnetic reversal, a supernova, and the flooding of the ocean surface by freshwater from an Arctic lake (Alvarez et al., 1980). The smoking gun discriminating between these hypotheses emerged when a set of traces discovered in the field overwhelmingly favored the meteorite impact hypothesis. These traces included deposits rich in iridium, an element rare on earth but common in meteors (Alvarez et al., 1980; Smit and Hertogen, 1980), deposits rich in impact ejecta (Bohor, 1990; Montanari et al., 1983), and the discovery of a large crater on the Yucatán Peninsula in Mexico (Hildebrand et al., 1991). As a result of these discoveries, the set of competing hypotheses for the extinction of the dinosaurs was heavily culled. More recently, novel alternative explanations for the mass extinction have emerged, among them the massive volcanic activity in the Deccan Traps in India (Chenet et al., 2009), and the search for a new smoking gun continues.

The reliance on smoking guns means that historical sciences do not work by testing predictions. A prediction specifies what would happen under a specific set of conditions, given a certain hypothesis, and is tested by bringing about this set of test conditions, something that cannot be done without experimental methods (Cleland, 2001). Without experimental methods, it is impossible to know if a prediction failed because it is wrong or because the set of conditions it specifies were not brought about. The possibility of false-negative results explains why failed predictions rarely lead to the rejection of a hypothesis in historical sciences (Cleland, 2011, 2002). Instead, predictions, when historical scientists make them, serve as tentative guides in the search for smoking guns. They are educated guesses, based upon background knowledge, about where in the field additional traces may be found and what form these traces may take. In fact, whether a smoking gun is discovered as a result of a prediction or is simply stumbled upon has little bearing on the acceptance of the hypothesis it supports (Cleland, 2011).

In the end, without direct access to the past, the capacity of historical scientists to learn things about the past hinges entirely on the discovery of smoking guns in the field. Like detectives, they must snoop around for incriminating traces in nature.

A Likelihood-Ratio View of the Search for a Smoking Gun

There are two key aspects to the search for smoking guns in historical sciences. The first aspect, discussed above, is that smoking guns are not manufactured experimentally but found in nature. The second aspect is that smoking guns discriminate *between* competing hypotheses. This is a crucial distinction that many archaeologists have failed to recognize.

The smoking-gun approach can be operationalized in terms of the likelihood ratio. Imagine that we have two rival hypotheses to explain a certain phenomenon. Let us call the first hypothesis H_1 and the second one H_2. To test the two hypotheses, we have a set of data, D, that we collected in the field.

The likelihood ratio is a way to compare the relative likelihood that each hypothesis explains the data. The likelihood ratio of H_1 and H_2 is the ratio between two quantities, $p(D \mid H_1)$ and $p(D \mid H_2)$. The first quantity, $p(D \mid H_1)$, is the probability of observing data D, assuming that H_1 is true. For example, what is the probability of rolling a 6 given that the die is fair? Conversely, $p(D \mid H_2)$ is the probability of observing data D, assuming that H_2 is true. For instance, what is the probability of rolling a 6 given that a die is loaded in such a way that a 6 is scored four times more likely than the other sides? The likelihood ratio is the ratio of the probabilities that the two hypotheses have generated the observed data:

$$(1.1) \qquad \text{likelihood ratio} = \frac{p(D \mid H_1)}{p(D \mid H_2)}.$$

Equation 1.1 shows that when D can account equally well for both H_1 and H_2, the likelihoods are equal, and the likelihood ratio is 1. A likelihood ratio of 1 thus means that D is not a smoking gun for either hypothesis. But D is a smoking gun for H_1 if the likelihood ratio is greater than 1, or it is a smoking gun for H_2 when the likelihood ratio is smaller than 1. This is the likelihood-ratio view of the smoking gun: a smoking gun is data that tip the likelihood ratio away from 1. And the farther away from 1 the likelihood ratio is tipped, the more smoke there is.

The point here is not that historical scientists should use the likelihood-ratio test as a statistical method. After all, it is not always feasible to assign a specific number to terms like $p(D \mid H_1)$ or $p(D \mid H_2)$, especially when our theories are verbal and our data are qualitative. Rather, the point is that the likelihood-ratio view of the search for smoking guns is a useful way to understand the logic that underlies how successful historical sciences work. The likelihood-ratio view emphasizes the importance of explicitly taking into consideration the different explanations that can reasonably account for the data at hand—something archaeologists rarely do. In fact, many archaeologists do not even think of their research program as a hypothesis-driven enterprise. Yet, archaeologists test hypotheses all the time: every component of an archaeological interpretation is a hypothesis that is vulnerable to testing (R. Gould and Watson, 1982; Schiffer, 1988). Every time we infer something from archaeological material, every time we construct a narrative of what happened in the past, every time we draw a conclusion, we have generated, tested, and accepted a hypothesis, even if implicitly. Looking at historical sciences through the lens of the likelihood-ratio test forces us to acknowledge that we are constantly testing hypotheses. But more importantly, it emphasizes the fact that a good smoking gun discriminates *between hypotheses*, instead of merely being consistent with *a* hypothesis.

Archaeologists Use the "Test of Consistency" to Test Hypotheses

In practice, the way archaeologists test hypotheses rarely bears any resemblance to the likelihood-ratio method (eq. 1.1). Rather, they settle on an explanation simply because it is *consistent* with the data. Given empirical data D, a working hypothesis H_1 successfully passes the test of consistency when

(1.2) $p(D \mid H_1) > 0,$

where, again, $p(D \mid H_1)$ is the probability of observing the data, assuming that hypothesis 1 is true. A $p(D \mid H_1)$ greater than 0 means that the hypothesis is consistent with the data, at least to a certain extent. The greater $p(D \mid H_1)$ is, the more consistent the hypothesis is thought to be.

A more sophisticated version of the test of consistency is based on the rejection of a null hypothesis using the p-value. The null hypothesis is a statistical model in which causality is absent, and the p-value represents the probability of obtaining the data observed (or more extreme observations), assuming that the null hypothesis is true, or $p(D \mid H_{null})$. It is not, as is often assumed, the probability that the null hypothesis is true, given the data, $p(H_{null} \mid D)$, nor is it the probability that the target hypothesis H_1 is true, $p(H_1 \mid D)$. The null-hypothesis version of the test of consistency looks like this: a hypothesis H_1 is consistent with empirical data D when

(1.3) $p(D \mid H_{null}) < \alpha,$

where α is the significance threshold, typically set to 0.05, below which most null hypotheses are rejected. According to equation 1.3, a hypothesis is deemed consistent with the data when the probability of the null model generating the data at hand is less than 5%.

At first glance, the null-hypothesis testing depicted in equation 1.3 looks like the testing of two competing hypotheses, H_1 and H_{null}. But the rejection of the null hypothesis using p-values does not discriminate between H_1 and H_{null}; it is concerned with only the null hypothesis. Notice that H_1 is absent from equation 1.3: the p-value is completely independent of H_1. This means that the p-value has little bearing on the epistemic value of H_1. Imagine that an archaeologist is analyzing two ceramic assemblages from two different cultural levels. The vessels coming from the older level vary a lot in shape and size; those coming from the younger layer all look similar. An archaeologist hypothesizes that the vessels from the first level were produced by the members of different households, while those from the second level were produced by craft specialists. The archaeologist analyzes the data and obtains a "significant" p-value: the variance of the first assemblage is significantly larger than the variance of the second. He concludes that the data confirm the craft specialist hypothesis. Maybe this conclusion is right. But who knows? In reality, the rise of elite craft specialists was never tested directly. You could replace the rise of craft specialists by any other explanation, including fanciful ones that involve an extraterrestrial civilization, and the p-value would not budge by one decimal.

The test of consistency is especially prevalent in narrative interpretations of the archaeological record. We find that our ideas about what makes humans tick (i.e., our theories and hypotheses) are supported empirically when, in some way or another, they can account for the data at hand. The research based on the test of consistency typically starts with a discussion of some theory (e.g., costly signaling theory), followed by an archaeological case study (the zooarchaeological record of the Archaic period in southern Ontario), a demonstration that the data are consistent with the theory (big animals were preferentially hunted), and an interpretation of the data in terms of the theory (Archaic hunters from southern Ontario were hunting for prestige). The research paper may end with a discussion of how useful the theory is to archaeological research, and its title may read something like "Theory X: A View from Location Y."

Note that the test of consistency can be applied at different scales. It can be used to test a single hypothesis (the metal grave goods in this burial are prestige goods) or complex sets of hypotheses (grave goods denote social status) or a whole theory (a complex system view of state societies). Thus, an archaeologist may very well be using the smoking-gun approach to discriminate between a set of hypotheses while at the same time using the test of consistency to select the theory from which the hypotheses were drawn.

The test of consistency is different from the search for smoking guns depicted in equation 1.1. The likelihood-ratio view of the search for smoking guns is that it is not the *absolute* capacity of a hypothesis to account for the data that matters but its capacity *relative* to other hypotheses. For instance, a quantity such as $p(D \mid H_1)$ does not mean much in and of itself. Instead, it becomes meaningful only when it is compared with $p(D \mid H_2)$. For instance, it is not enough to show that the rise of elite craft specialists is consistent with the data; that hypothesis also has to account for the data better than alternative explanations for variation in ceramic vessels.

By focusing archaeologists' attention on a single hypothesis at a time, by telling them that demonstrating consistency is enough for science to advance, the test of consistency has led them to ignore the underdetermination problem that plagues archaeology. Furthermore, this underdetermination problem is amplified by how vulnerable the test of consistency is to confirmatory bias.

THE TEST OF CONSISTENCY LEADS TO CONFIRMATORY-BIASED RESEARCH

Another problem with the test of consistency is that it opens the door to a confirmatory bias (Klayman and Ha, 1987; Nickerson, 1998; Oswald and Grosjean, 2004). In 1890 the American geologist Thomas C. Chamberlain published a paper in which he explained the problems that arise from accepting an explanation because it is consistent with the data and without paying attention to alternative explanations. Chamberlain compared the testing of a single hypothesis to the blinding love of a parent for an only child. This love puts the researcher in danger of "an unconscious selection and of magnifying of phenomena that fall into harmony with the theory and support it, and an unconscious neglect of phenomena that fail of coincidence" (93). In contrast, working with multiple hypotheses "neutralizes the partialities" (93) of our "emotional nature" (93) and "promotes thoroughness" (94). Chamberlain was describing, more than 120 years ago, what psychologists today call the confirmatory bias.

The confirmatory bias is a strong cognitive bias that affects each and every one of us in our daily lives. In science, it arises when researchers show a bias for evidence or for certain interpretations of the evidence that reinforce their own views of how the world ought to be. Confirmatory bias can creep in at every stage of the research process, from the collection of data to peer review (Hergovich, Schott, and Burger, 2010; Koehler, 1993; Mahoney, 1977; Mahoney and Kimper, 1976; Resch, Ernst, and Garrow, 2000).

The test of consistency provides no protection whatsoever against confirmatory bias. To the contrary, because it amounts to little more than interpreting the data in terms of some hypothesis, the test of consistency naturally leads researchers to seek out a confirmation of their ideas and selectively ignore the traces that disconfirm their ideas as well as the alternative ideas that are also supported by the data. This happens not out of dishonesty but because this is how our brains work unless restrained by scientific methods.

THE TEST OF CONSISTENCY MAKES VERBAL HYPOTHESES TOO EASY TO CONFIRM

The test of consistency also makes it too easy for hypotheses to be confirmed, a problem that is magnified when the theories and hypotheses tested are verbally and imprecisely described. The great German writer Goethe is mostly known for his literary work, but he was also deeply interested in natural sciences. In 1810 he published a treatise on the perception of colors, *Theory of Colors*. In it, he describes electricity as "nothing, a zero, a mere point, which, however, dwells in all apparent existences, and at the same time is the point of origin whence, on the slightest stimulus, a double appearance presents itself, an appearance which only manifests itself to vanish. The conditions under which this manifestation is excited are infinitely varied, according to the nature of particular bodies" (Goethe, 1970, 295).

Goethe's view of electricity is so imprecise and unclear that it can be made consistent with just about any kind of phenomenon (Chalmers, 2013). In several fields, including archaeology, theories and hypotheses are described verbally as opposed to mathematically. Although archaeological hypotheses are described in a more precise manner than Goethe's theory of electricity, they remain, because of their verbal nature, imprecise enough that they can be made consistent with empirical evidence very easily.

Take the idea that as foraging intensity increases, prey items become smaller. The terms that link the theory to the empirical world, "increases" and "become smaller," leave a lot of leeway for the idea to be confirmed. The idea is consistent with just about any dataset in which prey items decrease in size, by whatever amount. In contrast, if you were to translate the same hypothesis mathematically, and work out the math, you might find out that prey item sizes decrease with foraging intensity following a particular function (linear, exponential, etc.). Unlike the verbal version of the same hypothesis, the formal version is consistent, not with just any kind of decrease in prey item size, but only with a specific mode of decrease.

In contrast to the test of consistency, the smoking-gun approach depicted in equation 1.1 entails a detailed understanding of how the causal mechanisms hypothesized operate—it forces us to flesh out and articulate our ideas better. Take again the example of the hypothesis that the rise of elite craft specialists is reflected in a decrease in the variance in craft goods such as ceramics. In and of itself, the idea is straightforward to confirm: does the variance in craft goods in a region decrease as social complexity increases, yes or no? But testing the same idea in the context of a search for smoking guns is more difficult. The smoking-gun approach demands that we show that the elite craft specialist hypothesis accounts for the data *better than* the other processes that are known to decrease within-group cultural variation, such as social norms, conformist-biased transmission, or functional pressures. This requirement, in turn, demands a detailed mechanistic account of how craft specialization, social norms, conformist-biased transmission, and functional pressures work and how they vary in their outcomes. This is much harder to accomplish than merely showing that variance decreases with time.

HYPOTHESIS REJECTION DOES NOT REDEEM THE TEST OF CONSISTENCY

The test of consistency is sometimes defended on the basis that it allows us to reject hypotheses when they are inconsistent with the data. In principle, the test of consistency should indeed allow us to falsify hypotheses, but in reality, that does not happen often. This is because in historical sciences, the data that appear to falsify a historical hypothesis can often be explained away by invoking factors that are extraneous to the target hypothesis, such as confounding variables, measurement errors, and sampling errors (Cleland, 2001).

Invoking extraneous factors to protect an unsupported hypothesis may seem, at first glance, like bad science—an attempt to salvage a pet theory with ad hoc reasoning. But the rub is that historical scientists always face the very real threat of false results, since they cannot experimentally manipulate past conditions.

Historical records are subjected to information-destroying forces: false-negative results are necessarily commonplace. It is always possible that the traces that would normally support a hypothesis have yet to be discovered or have been obliterated from the surface of the earth. Thus, the *absence* of something is rarely enough to falsify a hypothesis.

But false-positive results are possible too. Observations are prone to error, and the trace that falsifies a hypothesis may be rejected on the basis of measurements or methodological grounds (McElreath, 2016). This is why even research questions that are simple and that should be, in principle, falsifiable are not easily rejected. For instance, the report of an early occupation of the site of Monte Verde, Chile (Dillehay, 1989), was, at face value, a serious blow to the Clovis-first hypothesis for the colonization of North America. But what should have been a swift death dragged into a decades-long protracted debate, with some archaeologists explaining away the data by questioning the validity of the radiocarbon dates obtained from the site, their association with the artifacts, as well as the field methods used during the excavation (Lynch, 1990; West, 1993).

What is more, many of the data that archaeologists use to rule out hypotheses are not direct empirical measurements, such as radiometric age estimates, but "second-degree" data, that is, data produced through middle-range theory, such as a social network reconstructed from raw-material sourcing or a population size inferred from the size of ceramic assemblages. These second-degree data are even more prone to measurement errors and vulnerable to critiques than first-degree observations.

The difficulty of rejecting hypotheses is an issue that all historical sciences face. And it is not one that the smoking-gun approach resolves. But the message is that hypothesis rejection in historical sciences is not an efficient process: it is, at best, slow and messy. The possibility that the test of consistency leads at times to the rejection of a hypothesis does not outweigh its many costs and does not justify it.

THE TEST OF CONSISTENCY LEADS TO WRONG RESULTS

The fact that an explanation is consistent with the data at hand has little epistemic bearing on its validity, especially if a confirmatory bias has influenced the research process, or if the explanation is verbal and imprecise.

In the best-case scenario, the test of consistency generates no new knowledge. An explanation that is consistent with the data is worth further investigation: it is a just-so story that has yet to be tested properly. And even the most clever, cogent, and insightful just-so story remains only that, a just-so story.

More likely, the test of consistency leads to wrong results. By shielding ideas from a true empirical test, the test of consistency lets us draw conclusions even when our data underdetermine their cause. It allows false beliefs, erroneous claims, and spurious chains of cause and effect to be maintained in a community for a very long time. In fact, the sparser and more imperfect the data are, the easier it becomes to confirm just about any hypothesis. Think of how the vast majority of observations we make in our daily lives are consistent with the idea that the earth is flat, and how easy it would be to pick, unconsciously, the evidence that allows us to maintain that belief. If the test of consistency leads easily to false beliefs, then, when combined with a confirmatory bias and an underdetermination problem, it most definitely will produce wrong beliefs.

A related consequence of letting beliefs fly under the radar of true empirical tests is that new ideas and theories are added to a field at a faster rate than they are eliminated. Without selection, new hypotheses are grafted to the existing pool of ideas rather than used to replace older, disproven ones. Under such a regime, ideas or theories disappear from the literature not because they have been found inadequate but because they have fallen out of fashion. This problem intensifies with the magnitude of the underdetermination problem. The larger the underdetermination problem is, the more room there is for the free play of the imagination, and the easier it is to come up with incompatible but equally consistent rival hypotheses (Turner, 2007). Thus, a symptom of a field that relies on the test of consistency and ignores the underdetermination problem is a balkanized theoretical landscape, composed of a vast range of unrelated, or even incompatible, theories and ideas.

In the end, the test of consistency is too weak to serve as the cornerstone of any scientific discipline. If archaeologists are to acquire valid knowledge about the past, it cannot be by accepting interpretations on the sole basis that they are consistent with the archaeological record.

Smoking Guns Must Be Found in Nature

Ultimately, the difficulties of doing historical sciences can all be traced back to the lack of direct access to the past. At first glance, it may seem like this problem can be circumvented by using simulations, models, ethnographic analogies, and experiments to make inferences about the past (Reid, Schiffer, and Rathje, 1975). But however useful, or even essential, to the scientific process these things are, they are not sources of smoking guns.

SIMULATIONS AND MODELS

Computer simulations and mathematical models behave a lot like experiments, and therefore, they may appear to be legitimate sources of smoking guns. For instance, the goal of simulations and models is to reveal the causal relationships among variables, keeping all other things equal. We find these causal relationships by varying, in a controlled manner, the parameters of the model, just like experimental scientists do in their laboratories. But whereas experimental scientists are investigating the empirical world, modelers are investigating the validity of their own thinking (Kokko, 2007; Wimsatt, 1987). We build models to help us verify the logic of an argument or to find the solution to a problem that is too complex for our limited primate brains. The results of simulations and models can tell us if a hypothesis is logically consistent and thus worth pursuing. They can generate new hypotheses and they may guide us in the field by directing our attention to things that we may have ignored otherwise. But in the end, simulations and models remain nothing more than sophisticated thought experiments. Their realm of action is confined to thoughts and theories and does not extend to the empirical world.

The results of simulations and models are only as secure as the assumptions built into them. Simulations and models are always simpler than reality. They may be simpler than reality by design: what makes models useful is that they allow us to trade realism for tractability. Or it may be by ignorance: we may have left important factors out of a simulation unintentionally, because we are unaware of their importance or even their existence. For example, early models of global climate indicated that nothing would be able to reverse a "snowball earth"—a global freeze of the earth's surface. But it would have been a mistake to treat the results of these models as a smoking gun that confirmed that the earth never experienced a global freeze, because these early models failed to include volcanic activity, which, it turns out, can emit enough carbon dioxide to produce a greenhouse effect and end a global freeze (Cleland, 2001).

ETHNOGRAPHIC ANALOGIES

The same line of reasoning applies to analogies. Historical scientists of all disciplines use contemporary analogues as surrogates for past, unobservable events, processes, and things, whether it is a young star that is used as a model for our sun in its early years, a modern lake for Pleistocene ones, or contemporary hunter-gatherers for Middle Paleolithic foragers.

Analogies have their place in historical science. In fact, it is impossible for a historical scientist to avoid analogical reasoning. Virtually every archaeological inference is based, somehow, on a contemporary analogue (Campbell, 1920, 1921; Chang, 1967; Gifford-Gonzalez, 1991; R. Gould and Watson, 1982; Mac-Cormac, 1976; Wylie, 1982, 1985; Yellen, 1977). The inference that "this artifact is a ceramic bowl" is analogical in nature. So is the naming of a bone specimen on the basis of its resemblance to modern bones (Gifford-Gonzalez, 1991). Less trivially, analogies are also useful as sources of novel testable hypotheses (Binford, 1966, 1967; Hempel, 1965; T. Murray and Walker, 1988). For example, the ethnographic record can make us aware of new alternative behaviors or direct our attention to different sources of evidence in the field. But analogies cannot, in and of themselves, serve as smoking guns, however consistent they are with archaeological evidence.

The ethnographic record is, by far, the main source of analogies in archaeology. Archaeologists have had a long and complicated relationship with ethnographic analogies, one that dates back to at least the nineteenth century (Lyman and O'Brien, 2001; Ormes, 1973), and the flaws and the virtues of ethnographic analogies have been debated at length over the years (e.g., Ascher, 1961; Chang, 1967; David and Kramer, 2001; R. Gould and Watson, 1982; Kelley and Hanen, 1988; Lyman and O'Brien, 2001; Wylie, 1982, 1985; Yellen, 1977).

As with all forms of inductive inferences, ethnographic analogies always run the risk of being wrong (David and Kramer, 2001; Kelley and Hanen, 1988). The researcher may have chosen the wrong analogue, either because he did not have enough information to choose between alternative analogues (Jeffares, 2010; Kelley and Hanen, 1988) or because he picked it on the basis of prejudice or sectarian opinion (Gee, 1999). Alternatively, he may have used the wrong analogue because he was trying to interpret the unknown in the light of the known, and the known is limited. The past phenomenon that we are trying to understand may very well fall outside the range of phenomena that can be observed in the ethnographic record (Freeman, 1968; Gee, 1999; R. Gould and Watson, 1982; Howell, 1968; Yellen, 1977). It would indeed be naive to expect past human societies to fall within the range of human behavior that happens to have been around in the twentieth and twenty-first centuries. Similarly, cross-cultural regularities that we see in the ethnographic record may disappear as cultures evolve in different directions and at different rates. This problem is compounded by our lack of a strong theory of human behavior, of the sort that fields such as biology, chemistry, and physics have and that would allow us to evaluate the robustness of the regularities identified in the ethnographic record (R. Gould and Watson, 1982). Given how variable human behavior is, the existence of an ethnographic analogue has little bearing on the value of an archaeological hypothesis that is based on the analogue (Binford, 1967; Hempel, 1965). Put simply, analogies are not a window into the past and cannot be used, in and of themselves, to discriminate between hypotheses.

EXPERIMENTAL ARCHAEOLOGY

Experimental archaeology and actualistic research suffer from similar limitations. Both types of research are useful in that they feed into the background knowledge from which historical hypotheses are derived. For instance, experiments can be conducted to understand how long bones break under specific types of mechanical stress. In this example, the investigator is engaged not in historical science but in true experimental science, as her object of study, the breakage of bones, is contemporary and directly observable.

Experimental and actualistic research are not sources of smoking guns, but they can, in certain conditions, help narrow down the number of competing hypotheses. This is especially true for low-level inferences about the physical world, whether it is identifying the species and sex of the individual from which a bone comes, whether a mark on a bone surface was left by a tooth or by a stone tool, what temperature a ceramic vessel was fired at, or whether a stone tool is a cutting or a pounding implement. In all these cases, our background knowledge derived from the study of contemporary analogues tells us that there are only a few alternative hypotheses that compete to explain the set of traces observed in the field, because the physical world is heavily constrained by biological factors (a gazelle cannot produce an offspring with a skull that looks like that of a hyena), physical factors (a ceramic fired in an open kiln will not vitrify), and mechanical ones (an axe does not make a good hammer). Thus, although based on analogical reasoning, these inferences about the past are "strongly warranted" (Gifford-Gonzalez, 1991). Experimental and actualistic studies, however, are much less warranted when we move away from low-level physical phenomena to the level of behavior, psychology, ecology, society, or culture. Despite decades of research in zooarchaeology, actualistic studies do not allow us to infer with any significant degree of confidence anything but the most proximal, immediate cause of a trace (e.g., a stone or a bone), and precious little about the intention or the strategy pursued by the actor, or the behavioral and ecological context of the trace (Gifford-Gonzalez, 1991; Lyman, 1994). This is because at these higher levels of explanation, the number of competing hypotheses that can explain the same set of traces is much larger.

In the end, the danger of simulations, models, ethnographic analogies, and experiments is that they give us the illusion that they can patch up an incomplete historical record; and they can lead us to assume the very things we should be trying to find out (Binford, 1968a; Clark, 1951; Freeman, 1968; R. Gould and Watson, 1982; T. Murray and Walker, 1988; Wylie, 1982). None of them offer any guarantee whatsoever of being accurate representations of the past, and because of that, they do not have the epistemological weight necessary to discriminate between competing hypotheses. They are simply sources of educated guesses about what the past may have looked like—sources of hypotheses that are interesting but that remain to be tested using field data.

NONEMPIRICAL VIRTUES

Scientists accept and reject hypotheses on the basis of not only how well they describe nature but also their nonempirical virtues (Fogelin, 2007; Glymour, 1984; T. Kuhn, 1962; Kukla, 1998; Psillos, 1999). Nonempirical virtues are global principles that apply to all fields of human inquiry and that are thought to break a tie between hypotheses that are equally supported empirically (Harris, 1994; Sober, 1988). All hypotheses are equal, but because of nonempirical virtues, some are more equal than others. For instance, the principle of parsimony stipulates that, all other things being equal, simpler hypotheses are better. Parsimony—along with predictive power, explanatory power, testability, lack of ad hoc features, capacity to generate new predictions, and compatibility with other theories—shapes scientific research in many ways.

Nonempirical virtues can operate behind the scenes, without the researchers even being aware of them. Nonempirical virtues help us separate the hypotheses that are reasonable and deserving of our attention from the ones that are bizarre and do not merit our time (Kukla, 2001). For example, no serious archaeologist would ever waste time testing a hypothesis that contradicted the laws of chemistry or that involved time-traveling astronauts. The nonempirical virtue of bizarre hypotheses is so low that they are ignored from the outset. Nonempirical virtues are also acting behind the scenes when we build statistical models. An infinite number of curves can be fitted to a series of data points, but we prefer simpler solutions over complicated ones, such as linear curves over complex polynomials (Forster and Sober, 1994; Kieseppä, 1997).

Nonempirical virtues are also used explicitly by researchers to break ties between hypotheses. This is where their value becomes uncertain. For one, the virtues are difficult to operationalize—defining what exactly "parsimony" or "explanatory power" means is easier said than done (Baker, 2011; Fraassen, 1980; Kukla, 2001; Sober, 1988; Turner, 2007). But more importantly, it is not always clear why nonempirical virtues should have any bearing on the value of hypotheses. What reasons do we have to think that nonempirical virtues are reliable indicators of truth, approximate truth, or likelihood? There are instances in which our theories tell us that some virtues are legitimate criteria by which to compare hypotheses. For example, in cladistics, the method of classifying taxonomic groups, what "parsimony" means is well understood, and its use is justified by what we know about evolution (Sober, 1988; Turner, 2007). In contrast, anthropology lacks a similar unifying theory of human behavior, culture, and society that would allow us to operationalize nonempirical virtues and justify their use.

We discriminate between hypotheses using nonempirical virtues mainly for practical reasons: a complex hypothesis, like a complex polynomial regression, is harder to defend than a simpler one. But that should not be mistaken for empirical support. "Parsimony" and "explanatory power" are not smoking guns and, like a simulation or an ethnographic analogue, do not have the epistemological weight of a smoking gun found in nature. To see why, consider how rapidly science would come to a grinding halt if the collection of new empirical data were to cease and scientists were left with only nonempirical virtues with which to discriminate between hypotheses. There is no way around it: the smoking guns that propel historical sciences must be found through fieldwork, in nature, and only there.

We Are at the Mercy of Nature

Historical scientists are at the mercy of nature. Until a smoking gun has been found, the historical science process outlined in figure 1.1 remains a stalled open loop. Historical scientists cannot manufacture smoking guns experimentally, and they cannot find them in computer simulations, mathematical models, analogies, or nonempirical virtues. Instead, they must find them in nature, through fieldwork.

In principle, solving the underdetermination problem should be easy: conduct more fieldwork or improve field techniques. And archaeologists do both. Every year, the portion of the earth that has been excavated expands, and new field techniques are constantly being developed, increasing the range of traces that we can detect. Indeed, many processes that were equifinal yesterday are perfectly distinguishable today, and the same may happen to processes that are, today, equifinal.

But being at the mercy of nature slows down the pace at which historical sciences can progress. The historical science process can remain a stalled open loop for a long period of time, especially if the smoking gun necessary to close the loop consists of minuscule or highly degraded traces that lie in wait, in nature, for the development of new technologies (Cleland, 2011, 2002). For instance, the organic-residue traces left inside ceramic vessels became observable only after the emergence of mass spectrometry methods in the 1950s and 1960s (Evershed, 2008). Alternatively, the smoking gun needed to close the loop may also be associated with rare events and thus unlikely to be discovered for sampling reasons. Or the smoking gun may exist but in a region of the world that is hard to reach.

More importantly, nothing guarantees that the smoking gun needed to resolve a scientific question will ever be discovered (Cleland, 2011). Finding a smoking gun, even if after a very long wait, is a best-case scenario. Nature, after all, does not care about our research interests. The traces about the past that have been left for us to discover do not have to be adequate for each and every one of our research questions. Although we are constantly improving our capacity to recover minute or degraded traces in the archaeological record, it may very well be the case that whatever smoking gun is required to solve a question either never entered the archaeological record or was completely erased from it. Historical records, such as the archaeological record, are constantly subjected to information degradation and loss, and when a particular smoking gun has been destroyed, it can never be found, and the historical science loop will never be closed, no matter how well we operationalize our research problems, no matter how much fieldwork we conduct, and no matter how much we improve our field techniques. Here, the source of underdetermination is the quality of the historical record *itself.*

The details of the picture of the scientific process painted in this chapter matter not. The historical research process depicted in figure 1.1, smoking guns, and the "test of multiple competing hypotheses" are nothing but heuristic devices that allow me to capture some of the issues with how archaeologists work. In practice, science is a messy and complex process, and archaeologists deploy a wide range of scientific methods in their research (Hegmon, 2003). What truly matters in the end is whether the empirical data at hand support our inferences *beyond a reasonable doubt*—that is, with a very high likelihood ratio. Until, if ever, a proper smoking gun has been found, historical scientists have to remain agnostic about what past cause explains the data. In this, they are not acting any differently from a jury that declares a person guilty beyond a reasonable doubt.

With this criminal justice analogy in mind, archaeologists can ask themselves whether a reasonable jury, when presented with the archaeological evidence at hand, would be convinced that the archaeologists' interpretation is right. During a criminal case, a jury is tasked with weighting two competing hypotheses: the defendant is guilty or he is innocent. They are asked to do so by using the rule of reasonable doubt. Reasonable doubt is the threshold of proof beyond which the evidence leaves no doubt that the defendant is guilty. The Supreme Court of Canada explained what the expression means in more detail:

> A reasonable doubt is not an imaginary or frivolous doubt. It must not be based upon sympathy or prejudice. Rather, it is based on reason and common sense. It is logically derived from the evidence or absence of evidence.
>
> Even if you believe the accused is probably guilty or likely guilty, that is not sufficient. In those circumstances you must give the benefit of the doubt to the accused and acquit because the Crown has failed to satisfy you of the guilt of the accused beyond a reasonable doubt.
>
> On the other hand you must remember that it is virtually impossible to prove anything to an absolute certainty and the Crown is not required to do so. Such a standard of proof is impossibly high.
>
> In short if, based upon the evidence before the court, you are sure that the accused committed the offense you should convict since this demonstrates that you are satisfied of his guilt beyond a reasonable doubt. (R. v. Lifchus, [1997] 3 S.C.R. 320, 1997 canlii 319 [SCC])

This definition of reasonable doubt reflects surprisingly well how historical sciences work. It warns the jury against the same biases that plague confirmatory research, like sympathy or prejudice. It acknowledges that we can never "prove" anything with absolute certainty—something that every scientist agrees on—but it also recognizes that consistency alone is not a sufficient criterion to find someone guilty. This resemblance is not accidental: criminal justice is, in a sense, a historical science, as it seeks to explain contemporary evidence (DNA, witness testimony, phone records) in terms of its past cause (who perpetrated the crime). The rule of reasonable doubt also speaks to the underdetermination problem. Evidence that fails to convince a reasonable person beyond reasonable doubt is evidence that underdetermines the guiltiness of the accused—it is evidence that lacks (perhaps quite literally) a smoking gun.

"Beyond reasonable doubt" is a high bar to pass. Why is it not enough to convict someone when we think that this person is guilty? Because we risk convicting an innocent, which, in our society, is considered worse than letting a guilty person walk free (Underwood, 2002). If juries were allowed to use the same test of consistency that archaeologists use, jails would be filled with innocent individuals and people would have no trust in the justice system. Of course, when historical scientists ignore the underdetermination problem and publish narratives of what they think happened in the past, no innocents are sent to jail. But their journals end up being filled with wrong results, and the confidence of the scientific community in their field is eroded.

Because we are at the mercy of nature, because of the quality of the archaeological record, there are research questions that we will never be able to answer beyond any reasonable doubts. The quality of the archaeological record constrains the range of research topics archaeologists can study. But nature is not always unmerciful and is more generous regarding certain kinds of research questions. Rather than waiting in vain for an unlikely smoking gun, archaeologists would do better to focus on those questions for which we can expect beyond-reasonable-doubt answers. Only in doing so can archaeologists unleash the full contributive value of archaeology for the sciences of human behavior. Before archaeologists can do that, however, they need a clear understanding of how the problem of underdetermination arises in archaeology.

2

The Sources of Underdetermination

The primary cause of underdetermination in archaeology is the quality of the archaeological record: whether or not we can discover the smoking gun necessary to resolve a research question depends in large part on how much information about the past has been preserved on the surface of the earth. There are several pathways by which a set of empirical observations can underdetermine a process, and if archaeologists are to shield their program from underdetermination, they need a clear understanding of what these pathways are. In this chapter I develop a general theory of underdetermination that links underdetermination to measurable aspects of the quality of data.

The Quality of Data

The underdetermination problem depends on at least four aspects of the quality of a dataset: its *scope*, its *sampling interval*, its *resolution*, and its *dimensionality* (table 2.1). These four aspects describe the properties of the analytical units that make up the dataset, as well as the relationships between these units. The aspects are not specific to archaeology—they can be used to describe the quality of any set of empirical observations in any discipline. For instance, paleontology has a rich literature discussing the quality of the fossil record in terms of its temporal scope (e.g., Martin, 1999; Schindel, 1982a, 1982b), its sampling interval (sometimes referred to as stratigraphic resolution or as paleontological, stratigraphic, or depositional completeness) (Behrensmeyer, Kidwell, and Gastaldo, 2000; Erwin, 2006; Kidwell and Flessa, 1996; Kidwell and Holland, 2002; Kowalewski, 1996; Kowalewski and Bambach, 2003; Kowalewski, Goodfriend, and Flessa, 1998; Martin, 1999; Schindel, 1982a, 1982b), and its resolution scale (alternatively called stratigraphic condensation,

TABLE 2.1. The different aspects of the quality of scientific data

Quality of data	Definition
Scope	The total amount of time and space that is encompassed in a dataset, i.e., the spatial and temporal width of the observation window
Sampling interval	The interval of time and space *between* each analytical unit, i.e., the sampling interval
Resolution	The amount of time and space that is represented *within* each analytical unit, i.e., the extent of time and space averaging
Dimensionality	The number of independent dimensions of an object of study measured

depositional resolution, duration, temporal resolution, microstratigraphic acuity) (Behrensmeyer and Chapman, 1993; Behrensmeyer, Kidwell, and Gastaldo, 2000; Erwin, 2006; Fürsich and Aberhan, 1990; Graham, 1993; Kidwell and Behrensmeyer, 1993; Kidwell and Flessa, 1996; Kowalewski, 1996; Kowalewski and Bambach, 2003; Kowalewski, Goodfriend, and Flessa, 1998; Martin, 1999; Olszewski, 1999; Schindel, 1982a, 1980; Walker and Bambach, 1971).

The first aspect of the quality of data, *scope*, refers to the total amount of space and time that is represented in a dataset. Take the example of the US census. The US government started taking a census of its population in 1790, and the most recent census was conducted in 2010. The current temporal scope of the US census dataset is thus 220 years (1790–2010). The spatial scope of the census is, today, the total area of the United States, or about 9.83 million km^2. In an archaeological context, the temporal scope of a dataset is the time span between the earliest and the latest analytical units, and its spatial scope denotes the surface area covered by the study.

Sampling interval denotes the interval of time or space that separates the analytical units. The temporal sampling interval of the US census data is ten years, since data about the country's population are collected once every decade. While the US census data themselves do not have a spatial sampling interval per se, a dataset comprised of census data from more than one country, say the United States and Australia, would have a spatial sampling interval equal to the distance between the United States and Australia. Archaeologically, sampling interval denotes the interval at which the archaeological record is sampled. For instance, in a study in which archaeological sites are the analytical units, the temporal sampling interval refers to the time gap between the ages of the sites. Similarly, the spatial sampling interval is the spatial distance between the sites.

Resolution refers to the amount of space and time that is represented *within* each individual data point in a dataset. The resolution of an analytical unit is thus a measure of the extent to which it collapses together events that took place at different points in time or in space. In other words, resolution refers to how *time averaged* and *space averaged* the units in a dataset are. The resolution of the US census is 24 hours, since the individuals who participated in the program filled out the census form on a specific day (April 1 in the most recent census). To give another example, the gross domestic product (GDP) of a country has a resolution of 1 year because it measures the market value of all the products and services produced in a country over a 1-year period; that is, it collapses into a single number 365 days of economic activity. Thus, the resolution of the GDP is, temporally, coarser than that of the US census. Spatially, the GDP collapses into a single data point all the economic activity that took place in the whole country and thus has the same spatial resolution as the US census. The resolution of an archaeological dataset is the extent to which the analytical units it comprises, such as assemblages, cultural levels, or sites, are constituted of material that comes from events that took place at different points in time and space. Thus, the resolution of an archaeological dataset is the finest temporal or spatial bin to which archaeological remains can be assigned, whether owing to the effect of taphonomic factors or by research design.

Note that neither *sampling interval* nor *resolution* is synonymous with *hierarchical level*. Hierarchical level refers to the level at which the analytical units are constructed. Every phenomenon in nature can be decomposed into various levels that are embedded into each other in a hierarchical fashion. For example, biologists may study life at the level of species interaction (ecology), the behavior of individual members of a species (ethology, behavioral ecology), the body of these individuals (anatomy), the internal functioning of the organs that make up the body (physiology), the cells that make up these organs (cell biology), the internal functioning of these cells (biochemistry), or the genes (genetics). Similarly, archaeologists study the human past at various levels. An archaeologist may study the archaeological record at the scale of a large region (the southeastern United States), a small region (the lower Mississippi delta), an individual archaeological site (the Winterville site, Mississippi), an activity area within a site (mound B at Winterville), or the features and objects within an activity area (burial 3) (Brain, 1969). Both sampling interval and resolution may covary with hierarchical level. For instance, the spatial resolution of a dataset in which the analytical units have been constructed at the hierarchical level of the feature will likely be finer than that of a dataset in which the units are archaeological sites. Likewise, the temporal sampling interval of archaeological data may be shorter at the level of individual artifacts within an occupation layer than at the level of archaeological sites. But different datasets constructed at the same hierarchical level can have very different sampling intervals and resolution.

Finally, *dimensionality* refers to the number of independent dimensions of an object that have been measured. Thus, dimensionality corresponds to the number of independent variables in a dataset. In the case of the US census, the object of study is the resident population of the United States. The independent dimensions of each resident that are measured include name, sex, age, date of birth, race, homeownership status, and residence location—a total of seven dimensions. In the case of a lithic database, the dimensionality may include many variables, such as the spatial location of the artifacts, their age, their raw-material type, as well as dozens of morphometric attributes (Andrefsky, 2005).

How scope, sampling interval, resolution, and dimensionality are measured depends on the hierarchical level at which the analytical units are constructed. In one dataset, the spatial sampling interval may refer to the horizontal distance between flakes in a lithic scatter, whereas in another dataset, it may refer to the

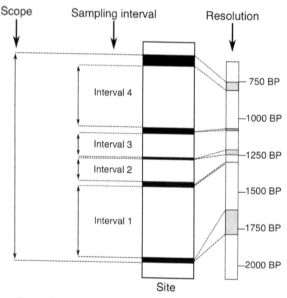

FIGURE 2.1: Temporal scope, temporal sampling interval, and temporal resolution in archaeological context. The scope of a dataset comprising cultural levels at a stratified site (dark areas) is the interval of time between the earliest and latest cultural levels. The temporal sampling interval is the distribution of time intervals between the layers, and the resolution is the amount of time represented within individual levels. (Adapted from Behrensmeyer and Hook 1992.)

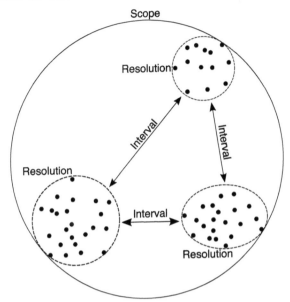

FIGURE 2.2: Spatial scope, spatial sampling interval, and spatial resolution in archaeological context. The figure represents an archaeological site with three concentrations of lithic artifacts. Assuming that the data are analyzed at the hierarchical level of the lithic concentration, the spatial scope represents the total area of the archaeological site; the sampling interval is the pairwise distance between the lithic concentrations; and the resolution is the area of the individual concentrations.

distance between the households of a village. Figure 2.1 shows how the temporal scope, sampling interval, and resolution are measured when the analytical units are cultural levels in a stratified site, and figure 2.2 illustrates how the spatial quality of a dataset is measured at the level of lithic concentrations.

In the end, the quality of data is commensurate with the amount of information a dataset contains: scope, sampling interval, resolution, and dimensionality are metrics that capture how much information a dataset contains. A dataset with a smaller scope, with larger intervals, coarser resolution, and fewer dimensions contains less information about the empirical world than a dataset with a larger scope, smaller intervals, finer resolution, and more dimensions. The less information contained in a dataset, the less likely it is to contain a smoking gun.

Scope and Underdetermination

The scope over which we observe nature limits the range of phenomena we can detect. Every phenomenon unfolds at a certain rate, both in time and in space. A population of bacteria can grow and double in size within a few minutes inside a Petri dish. A useful genetic mutation can sweep through an animal population and spread over a whole continent in just a few decades. Continents are moving away from each other at a rate of a few centimeters per year, whereas the disks of spiral galaxies form over billions of years and can stretch hundreds of thousands of light-years across.

The rate at which a phenomenon unfolds determines the scope over which we must take measurements in order to detect and study it (Bailey, 1981, 1983; Frankel, 1988; Stern, 1993). To determine the growth rate of a population of bacteria, one has to measure its size at least twice within the few hours the population needs to reach the carrying capacity of the Petri dish. The process of bacterial growth happens over an hourly scale, and so it can be studied only with a dataset that has a scope of a few hours. Similarly, when you spend a day at the beach, you notice the processes that operate over time scales that are shorter than a day and over spatial scales that are smaller than the spatial limits of your senses. These include the turnover of people, the movement of the tide, and, if you are unlucky, your skin burning. You notice these processes because they operate within the scope of your observations. Conversely, you will fail to notice the processes that operate over time scales that are longer than a day, like the erosion of the beach or tectonic drift. But visit the beach multiple times over the course of several years, and you may detect the erosion of the beach. Or zoom out from the vantage point of your beach chair in Guanabara Bay, Rio de Janeiro, to a satellite view of the planet, and you may notice that the coast-lines of South America and Africa have complementary shapes, suggesting that the two continents were once part of a single continental structure. For any given process, there is a minimum limit to the scope of empirical observation below which it will remain invisible. This is true even though the processes are operating continuously, like tectonic drift. It is the rate at which a process oper-ates, combined with the precision of our measuring instruments, that sets the scope necessary to detect and study it.

The scope of our observations also affects our capacity to observe the causal relationships between variables (Bailey, 1981, 1983). At the scale of our solar system, the effect of the finite speed of light is trivial and can be treated as a constant. But at an intergalactic scale, however, the effect of the speed of light becomes an important causal variable that needs to be accounted for when interpreting astronomical data (Bailey, 1981, 107). Similarly, when a pri-matologist observes a shift in the diet of chimpanzees over the course of a sin-gle week, she can exclude natural selection as a candidate explanation for the change. Natural selection, however, could cause a change in diet over scales of decades or centuries. The same goes for the role of climatic fluctuations on human behavior, which can be treated as a constant by an ethnographer spending a year at a field site but needs to be taken into account by an archaeologist as a possible driver of cultural change (Bailey, 1981).

The fact that the temporal scope of the archaeological record is much longer than the scope of the ethnographic record means that archaeologists may be able to detect processes and causal relationships that are invisible to ethnographers, a possibility that is discussed in more detail in chapter 7.

Sampling Interval and Underdetermination

One of the least recognized sources of underdetermination in archaeology is the discrepancy between the sampling interval of archaeological data and the scale over which the processes we want to study operate. A dataset with intervals that are too large undersamples the phenomena of interest and, as a result, lacks the smoking gun necessary to study it.

Imagine a stratified site with two cultural levels separated by 500 years. Two ceramic styles, *Red* and *Black*, are present in the two levels. In the older level, *Red* ceramic accounts for 90% of the ceramic assemblage, and *Black* ceramic accounts for the remaining 10%. In the younger level, the popularity of the ceramic style is reversed: *Red* accounts for 15% of the ceramic assemblage, and *Black* for 85%. (Throughout the book I use fictional archaeological cases instead of real ones. These simplified toy models allow me to home in on the relevant aspects of archaeological practice without getting lost in the details, as well as to avoid criticizing specific studies and authors.)

Different archaeologists may have different interpretations for this archaeological sequence, especially if they have different theoretical allegiances. One archaeologist may argue, for instance, that the reversal in the frequency of the styles is the product of agency, driven by changes in the social structure of the group. Another archaeologist may see in the data the rise of a class of craft specialists that correlates with the emergence of complex societies. An evolutionary archaeologist might find that the data confirm his belief that prestige-biased social transmission is the major source of cultural change. The problem with these interpretations is that agency, economic specialization, and prestige-biased social learning can all generate changes in material culture in less than 500 years. Because they operate over time scales that are shorter than the sampling interval of the data, they can all be made consistent with the data. Which one of these processes, if any, explains the data? Your guess is as good as any. This example also illustrates how sampling interval and resolution are independent sources of underdetermination: the data would still underdetermine their cause even if the two cultural levels had fine-grained, Pompeii-like resolution.

Sampling intervals lead to an underdetermination problem when they are so large that they miss the smoking gun necessary to discriminate between competing hypotheses. Consider figure 2.3. It shows how three different processes, A, B, and C, can increase the relative frequency of a ceramic style. The three processes lead to a linear, logistic, and stepwise increase function, respectively, but all lead, ultimately, to the same outcome: an increase in the frequency of a ceramic style from 20% to 80%, between time t_1 and t_2. In principle, the three processes could be discriminated by measuring the frequency of the style multiple times between t_1 and t_2. But data points with a sampling interval of $\Delta_t = t_2 - t_1$ or longer underdetermine the mechanism by which the ceramic style increased in frequency: the data cannot discriminate between A, B, and C.

Sampling intervals also lead to an underdetermination problem when they are so large that they make it impossible for us to distinguish signal from noise in the data. "Signal" here refers to the statistical pattern of interest. For instance, an ornithologist may want to know if there is a difference between the average body size of males and females in a species of birds. In this case, the relationship between body size and sex is the signal of interest. The "noise" is the variation in body size that is due to all the factors other than sex.

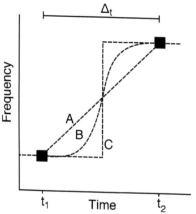

FIGURE 2.3: Sampling interval as a source of underdetermination. The relative frequency of a cultural trait (y-axis) changes over time (x-axis). Two archaeological assemblages have been recovered, representing material deposited at time t_1 and t_2 (black squares). The data collected from these two assemblages thus have a sampling interval of $t_2 - t_1 = \Delta_t$. Three different processes, A, B, and C, generate changes in the frequency of a cultural trait over time scales shorter than Δ_t. As a result, the processes are underdetermined by observations made at a sampling interval of Δ_t or more.

These factors include measurement errors, diet quality, or genetic variation. The noise can be such that it buries the signal: the distributions of males and females may overlap, so that many males are smaller than many females, and vice versa. This noise creates a problem for the ornithologist: if her sample size is too small, she may conclude that the females are larger than the males, whereas in reality the males are larger, on average, than the females. Detecting a signal amid a noisy background is perhaps the most important challenge in science. It is the reason why we have statistics, and why we spend so much time fretting about sample size. In fact, the terms "statistically significant" and "statistically nonsignificant" are used to differentiate the interesting (i.e., male and female body sizes are drawn from different populations) from the uninteresting (from the point of view of the research question) variation (i.e., the null hypothesis that the body sizes of males and females are drawn from the same population).

Large sampling intervals can mask the difference between significant and nonsignificant variation in our data, especially when the sampling intervals are larger than the scale over which the nonsignificant variation takes place. Imagine that the subsistence pattern of a group of foragers includes both marine and terrestrial resources. The diet of the group, however, changes with the seasons: it focuses on marine resources during the summer and on terrestrial food in the winter. This shift in diet is visible archaeologically, because the group produces mostly fishhooks in the summer and mostly projectile points in the winter. Now imagine that an archaeologist has excavated two assemblages, one representing a winter camp, and the other a summer camp, and that 200 years separate the two assemblages. The archaeologist may conclude that the subsistence pattern of the group has changed significantly during these 200 years, from a yearlong use of terrestrial resources to a yearlong use of marine resources. This, of course, would be wrong: the archaeologist mistook the noise (the seasonal variation in a group's diet—"nonsignificant" variation in terms of the overall subsistence pattern) for the signal (a change in subsistence pattern). The data underdetermine both the hypothesis that the subsistence pattern has remained the same and the hypothesis that it has changed because the sampling interval is longer than the scale over which the "nonsignificant" seasonal variation takes place.

Again, it is worth emphasizing the difference between sampling interval and hierarchical level. Whereas a dataset will underdetermine the processes that operate over a shorter time scale than their sampling interval, it may nonetheless be good enough to infer processes that operate at a lower hierarchical level. For example, Gregor Mendel discovered how genetic inheritance works using observations made at the level of individual phenotypes. By crossing varieties of peas and observing changes in their color from one generation to the next, he was able to show that an offspring's phenotype is jointly determined by two particles (i.e., genes) that it inherited from its parents, rejecting, by the same token, the prevailing model of blending inheritance according to which the parents' traits are averaged in the offspring. Mendel, however, would not have been able to infer the generation-scale process of genetic inheritance had there been a time interval of, say, 1000 generations between his observations.

Resolution and Underdetermination

Whereas overly large sampling intervals lead to underdetermination because they undersample the object of study, units with too large resolution do so because they oversample it. Units with overly large resolution collapse too much information into single units, burying the smoking gun within them.

Any given dataset will underdetermine the processes that operate over temporal or spatial scales that are shorter than its resolution. Let us go back to the example of the three processes that lead to an increase in the relative frequency of a ceramic style between time t_1 and t_2 (fig. 2.3). For the sake of simplicity, let us assume that ceramic material is deposited at a site at a constant rate between t_1 and t_2. Now, imagine that all the material deposited between t_1 and t_2 is lumped into one archaeological assemblage (fig. 2.4). Thus, the assemblage has a resolution of $t_2 - t_1 = \Delta_t$. Far from a Pompeii-like snapshot, the assemblage collapses all the information about the frequency of the trait at every point in time between t_1 and t_2 into one single number: the average frequency of the trait over that time period (the black square in fig. 2.4). The process that influenced the frequency of the cultural trait between t_1 and t_2 cannot be identified in the lumped assemblage. In the case shown in figure 2.4, the three idealized processes are symmetrical around $t_{1.5}$ and will all leave the exact same archaeological signature when averaged over t_1 and t_2. Data with resolution Δ_t thus underdetermine the processes that operate over shorter time scales than Δ_t.

Like sampling interval, resolution can also blur the signal and the noise and create the false appearance of change. The same phenomenon can leave different empirical signatures depending on the resolution of the data. Think back to the example of a population that focuses on marine resources during the summer and on terrestrial ones in the winter. The ratio of hooks to points will be different depending on how many summers and winters are represented in one assemblage. For instance, the hook-to-point ratio will increase as one goes from an assemblage that mixes the material deposited during one

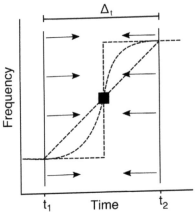

FIGURE 2.4. Resolution can be a source of underdetermination. The relative frequency of a cultural trait (y-axis) changes over time (x-axis). Coarse resolution can make competing processes indistinguishable. The frequency of the trait in an assemblage with resolution $t_2 - t_1 = \Delta_t$ (black square) looks the same whether it has increased following a linear, logistic, or stepwise function.

summer and one winter (resolution scale of 1 year), to one that mixes two summers and a winter (a resolution scale of 1.5 years), or to three summers and two winters (resolution scale of 2.5 years). An archaeologist looking at a sequence of assemblages with different resolutions may conclude, wrongly, that the subsistence pattern of the population varied over time. Data with overly large resolution can thus underdetermine the hypotheses that different assemblages are drawn from the same population or from significantly different ones.

Dimensionality and Underdetermination

Another important source of underdetermination in archaeology is the missing of one or more dimensions in a dataset. Identifying a causal relationship between two variables almost always necessitates the control of covariates. Covariates are variables that are correlated with the variable of interest and whose effect can be mistaken for that of another predictor, leading to a false-positive or false-negative result. For instance, if we want to determine whether or not variation in wealth drives variation in house size in a settlement, we need to control for extraneous variables that can also affect house size, such as number of family members. A dataset that contains only the variables "house size" and "wealth" will underdetermine the cause of variation in house size, and a researcher may mistakenly conclude that wealth predicts house size even when it does not. Controlling for covariates is particularly important when studying complex phenomena such has human behavior, culture, and society, which have long and intricate chains of cause and effect.

Scientists have several strategies to deal with covariates. First, if they are experimental scientists, they may conduct controlled laboratory experiments. Laboratories are, by design, simplified versions of the real world and allow a researcher to eliminate from the outset a large number of covariates. As for the remaining covariates, they are experimentally kept constant while the target variable is varied.

The second strategy consists in using randomized controlled trial experiments and natural experiments. Randomized controlled experiments take place outside the laboratory, but they approximate laboratory conditions by assigning subjects randomly to different experimental conditions, so that the control group and the experimental group are similar in every respect except for the target variable. This ensures that any effect detected can be attributed to the target variable and only to it. Natural experiments work in similar ways. A natural experiment is one in which subjects have been exposed to the experimental or the control conditions "naturally," that is, beyond the control of the investigators but in a way that, again, approximates the conditions of a controlled experiment (Dunning, 2012). For instance, adoption studies of twins who are genetically identical but raised in different families have offered researchers a way to study the effect of environment on mental disorder and cognitive ability, controlling for the effect of genes.

The third strategy is to use statistics. Many scientists, especially social scientists, cannot conduct laboratory or controlled experiments, for ethical and practical reasons. Instead, they conduct observational studies. An observational study is one in which a researcher draws inferences from observing a population but without having control over the independent variables (Rosenbaum, 2002). As a result, the investigator has no way of knowing for sure that the different groups she compares are truly comparable (i.e., similar in every way but the target variable). To alleviate this problem, observational scientists try to mute the effect of covariates by controlling for them statistically. Statistical tools like multilinear regression seek to isolate the effect of a target variable by keeping the effects of covariates constant. When a paper stipulates that having siblings decreases stress levels in adulthood, *controlling* for the effect of sex, education, and parenting style of the parents, the author of the paper implies that the presence of siblings, sex, education, and parenting style of the parents are covariates and that a dataset that excludes them could underdetermine their effects. The author, however, avoided the underdetermination problem by collecting data that had enough dimensions to partition, statistically, the relative effects of these covariates. The statistical strategy of controlling covariates is effective only inasmuch as datasets have a sufficient number of dimensions. This is why social scientists have long made it a priority to collect multidimensional datasets that include as many potential covariates as possible. And even then, there is always the possibility that an unknown covariate, and not the independent variable, is responsible for the result, hence the proverbial warning that correlation does not equal causation.

Archaeologists, like other historical scientists, have used this third strategy, statistics, to shield their hypotheses from false results. This is not surprising—they cannot, after all, assign dead people to different test conditions. But how effective this strategy is at eliminating the effect of covariates depends on which dimensions can be measured in the archaeological record.

Yet, archaeologists could also use the second strategy. They could, by analyzing large time-averaged and space-averaged samples, create conditions that are analogous to controlled experiments, a possibility that is discussed in more detail in chapter 7.

The Magnitude of Underdetermination Is Relative

The magnitude of the underdetermination problem is relative to the quality of the data at hand and the particular process studied. The larger the sampling interval and the coarser the resolution of our data, relative to the rate at which a process operates, the more likely the data are to underdetermine the process. Similarly, complex processes with many factors and with complex chains of causes and effects require data with more dimensions in order to be studied. Thus, a dataset may be high quality with respect to a certain process but low quality with respect to another one.

Take the measurements made on ice cores from the Greenland ice sheet. Scientists have been able to link the isotopic composition of the ice cores to global atmospheric temperatures over tens of thousands of years of earth's history. Because a new layer of ice is deposited every year, scientists can associate individual layers of ice with specific years. The ice-core data thus track fluctuations in global temperature with a sampling interval and a resolution of 1 year. The scope of ice-core data is hundreds of thousands of years long—in the case of the North Greenland Ice Core Project, for instance, the scope is about 123,000 years (Svensson et al., 2005).

A wide range of processes affect the earth's atmosphere (Ahrens, 1998; Weisberg, 1981). Some of them unfold on a minute-to-minute basis, such as thermals—bubbles of hot air that rise from the surface of the earth. Other processes operate on an hour-to-hour scale, like the 24-hour rotation of the earth on its axis, which causes temperatures to rise in the morning, peak in the afternoon, and decrease in the evening. On a monthly time scale, the day-to-day changes in the position of the earth relative to the sun, as the earth completes its 365-day orbit around the sun, generate seasons. Other processes act over much longer time scales. For instance, sunspot activity has an 11-year cycle, the precession of the earth's axis of rotation occurs on a 26,000-year cycle, and the shape of the earth's orbit varies, from nearly circular to slightly more elliptical and back again, on a 100,000-year cycle. Similarly, the amount and the distribution of landmasses on the planet affect worldwide temperature over millions of years. Typically, these processes are divided into weather and climate, depending on the time scale over which they operate (Ahrens, 1998). Weather processes refer to the state of the atmosphere at a specific point in time and space and vary from a minute-to-minute to a day-to-day basis. Climate processes are averaged over longer periods of time and larger spatial areas.

The processes that the ice-core data underdetermine are dictated by the quality of the data. With a sampling interval and resolution of 1 year, the ice-core data underdetermine processes that operate over less than a year, such as daily variation in temperature. If the ice-core data were to have a resolution and a sampling interval of a day, however, it would allow us to study processes that operate over days, weeks, and months. It would still, however, underdetermine the processes that unfold over less than a day, like hourly variation in temperature.

The ice-core data are thus a low-quality record of past weather, as the 1-year sampling interval and resolution of the data are larger than the time scale over which weather unfolds. In addition, the global spatial scale of the ice-core record is much larger than the spatial scale over which weather changes. As a result, the ice-core data underdetermines local weather. For instance, it lacks the smoking gun necessary to assess whether it rained or not on the evening of November 15, 56,703 BP, over the location where Tempe, Arizona, stands today. Not only is the information about this specific weather event absent from the ice-core data, but it has been, as far as we know, lost forever.

On the other hand, the ice-core data are a high-quality record of past global climate. The sampling interval and the resolution of the ice-core data are equal to or shorter than the scale over which many climatic processes unfold. The global spatial resolution may be too large to study regional climate (e.g., the climate of California), but it is equal to the spatial scale over which global climate varies. Finally, while ice-core data—with two dimensions, age and average global temperature—may not be rich enough to allow us to study certain processes, the information is sufficiently rich to reveal several aspects of past climates, such as fluctuations in global surface temperatures.

But what about the processes that operate over time scales that are *larger* than the sampling interval or the resolution of the data? Do the ice-core data, with their 1-year sampling interval and resolution, underdetermine the processes that operate over time scales longer than 1 year? That depends on the scope of the dataset. A dataset with a 1-year resolution, a 1-year sampling interval, and a scope of 100 years can potentially determine processes that operate over time scales ranging from 1 year to 100 years. Similarly, a dataset with a 1-year resolution, a 1-year sampling interval, but with a scope of 1000 years can determine processes that unfold over time scales ranging from 1 year to 1000 years.

If the processes are ordered on a line according to the temporal scale over which they operate, the range of processes that can be studied with a dataset has a lower bound and an upper bound (fig. 2.5). The lower bound is defined by the sampling interval or the resolution of the dataset—whichever is longer. Processes that operate over time scales shorter than this boundary are underdetermined. The upper bound is defined by the scope of the dataset. Processes that operate over time scales that are longer than the scope of the dataset are also underdetermined. In the example shown in figure 2.5, the dataset has a sampling interval and a resolution of 1 year and a scope of 10,000 years. The dataset thus has the necessary sampling interval, resolution, and scope to study processes that generate variation over time scales that range from 1 year to 10,000 years. The same logic also applies to the spatial dimension of the dataset.

The number of dimensions contained in the dataset further reduces the set of processes that can be studied. In the example shown in figure 2.5, while the dataset has the sampling interval, the resolution, and the scope necessary to study processes that operate over 1–10,000 years, it may not have the dimensionality required to identify all of them. Covariates that also operate over time scales of 1–10,000 years are, effectively, competing hypotheses that need to be ruled out before the process of interest can be determined. Unless the dataset contains measurements of the covariates, the object of study will be underdetermined by the data.

The covariates that operate over shorter time scales than the sampling interval or resolution (<1 year in fig. 2.5) can be controlled for statistically or by averaging over them. Averaging over covariates can be achieved by analyzing either low-resolution data or large samples. For example, in the case of the ice-core climatic record, the effect of rapid and localized weather events, such as storms, is averaged over in the ice-core layer. In other words, the difference

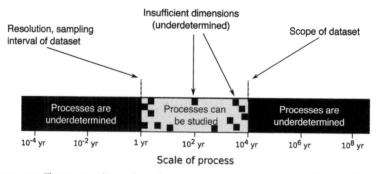

FIGURE 2.5: The processes that can be studied are those that operate over time scales that fall between the resolution or the sampling interval of the dataset (whichever is longer) and the scope of the dataset. Of the processes that fall between these two boundaries, some (black squares) will also be underdetermined because the dataset lacks the dimensions necessary to control for competing covariates.

between global temperatures as estimated from two different layers of ice cannot be due to the effect of a rapid, short-term, local event. Similarly, large samples can also allow us to mute the effect of covariates that operate over short time scales by averaging over them.

LONGER SAMPLING INTERVAL AND LOWER RESOLUTION UNDERDETERMINE MORE PROCESSES

The number of competing processes that can potentially explain any pattern in our data increases with the sampling interval or the resolution of the data. Imagine that you meet one of your friends in San Francisco one morning and that he calls you 10 hours later from New York. You can safely assume that, after your encounter, your friend boarded an airplane, because only an airplane can move an individual over thousands of miles in just a few hours. In other words, there is only one hypothesis for how your friend got from San Francisco to New York. But what if your friend, instead of calling you the same day, had called you a week later? Now you cannot be entirely sure that he took a plane, because there are other means of transportation that could allow him to get from San Francisco to New York in a week or less. Sure, chances are that he boarded an airplane, but you cannot rule out the possibility that he drove his car across the country or that he took a bus or even a train. As the interval of time between the two moments when you observe your friend's location increases from 10 hours to a week, the number of competing hypotheses you have to consider increases from 1 to 4. Had your friend called you 6 months later, you would also have to contend with a fifth hypothesis: that he crossed the country on his bicycle.

This example is trivial, but the exact same problem arises in archaeology. The set of processes that can explain a change in an archaeological variable over an interval of 100 years comprises all the processes that can lead to a change within 1 year, plus all the processes that can lead to a change within 2 years, plus those that operate within 3 years, and so on. The potential for underdetermination thus increases with sampling interval and resolution. A scientist who insists on interpreting a pattern that her data actually underdetermines is picking one hypothesis among the n candidate hypotheses that can equally explain the data. The chance that this scientist picked the correct hypothesis is $1/n$. As the sampling interval or the resolution of the data increases, so does n, so that $1/n$ quickly converges to 0. At that point, any interpretations drawn from the data are most likely wrong, for the same reason that any given lottery ticket is unlikely to be the winning one.

Reducing the magnitude of the underdetermination problem can be accomplished by improving the quality of the data. But for the historical sciences, there is only so much that can be done to improve the data quality. The technological breakthrough that will solve an underdetermination problem may not come until decades down the road, if ever. And, being at the mercy of nature, it is always possible that data that exist in nature will irremediably and forever underdetermine certain causes. When the quality of the data cannot be further improved, there is only one thing left to do: abandon, purely and simply, the study of the processes that are underdetermined. Instead, we should focus on the processes that the data do not underdetermine—those that operate over temporal and spatial scales that are similar to those of the data and whose complexity is matched by the data's dimensionality.

3

The Forces That Shape the Quality of the Archaeological Record, I: The Mixing of Archaeological Data

The archaeological record is shaped by many different forces. These forces operate in different ways—some of them act locally, others globally. Some operate episodically, some uniformly, while others increase in amplitude with time. But however they operate, their result is the same: a loss of information about the past.

These forces can be divided into two broad categories based on how they destroy information: those that *mix* archaeological material and information, and those that lead to their *loss*. I focus on the former in this chapter and on the latter in the next.

In both chapters I use models and simulations that build upon the rich literature archaeologists have produced over the years on the issues of taphonomy and site formation processes. But whereas much of this literature seeks to identify how the effects of mixing and loss can be controlled for, my focus is on their impact on the sampling interval, the resolution, and the dimensionality of data. Although the list of forces of mixing and loss reviewed here is not exhaustive, when these processes are considered together it is difficult to maintain the belief that the archaeological record is amenable to the same explanatory principles as the ethnographic record.

The Causes of Mixing of Archaeological Data

Three classes of forces mix archaeological remains: (1) depositional processes, (2) disturbance processes, and (3) analytical processes.

DEPOSITIONAL PROCESSES

Archaeological materials are mixed during deposition when they accumulate more rapidly than the matrix surrounding them does. Thus, the extent to which depositional process leads to mixing depends on both cultural and geological factors—the first determining how fast cultural material is deposited, and the second regulating how fast the matrix surrounding them accumulates.

Cultural Depositional Processes

Discard   The discard of objects by ancient people is the primary way by which archaeological assemblages form. It is also the first moment when material remains are mixed. The research on cultural discard, much of which is conducted under the umbrella of "accumulation research" (e.g., Gallivan, 2002; Mills, 1989; Schiffer, 1974, 1975, 1976, 1987; Shott, 1989a, 1989b, 2004; Sullivan, 2008a; Surovell, 2009; Varien and Mills, 1997; Varien and Ortman, 2005; Varien and Potter, 1997), is concerned with understanding how material objects leave the *systemic context* (the context in which artifacts are in a behavioral system) and accumulate in the *archaeological context* (the context in which artifacts interact only with the natural environment) (Schiffer, 1972).

One of the fundamental lessons learned from accumulation research is that the amount of archaeological material in an assemblage depends on the length of occupation of a site, the size of the group that occupies the site, and the rate at which objects are discarded (Baumhoff and Heizer, 1959; Cook, 1972a, 1972b; N. Nelson, 1909; Varien and Mills, 1997). The relationships between these variables can be described in the form of a discard equation (David, 1972; Schiffer, 1974, 1975, 1976, 1987; Surovell, 2009; Varien and Mills, 1997; Varien and Potter, 1997), the most influential of which being Schiffer's:

$$(3.1) \qquad d_t = \frac{S}{L} t,$$

where d_t is the total number of an artifact type discarded in a settlement during a period of time t, S is the average number of items of that type in use by the group at any given time, and L is the mean artifact use-life (i.e., the average length of time that an artifact of a specific artifact type is in use). This model of discard is concerned only with the discard of one type of object, and as such it is a simplification of Ammerman and Feldman's (1974) model of assemblage formation that deals with the simultaneous discard of multiple types (Shott, 2006).

Solving Shiffer's discard equation for t, the duration of discard (Varien and Mills, 1997), we get

(3.2) $$t = \frac{d_t L}{S}.$$

This form of the discard equation has been applied to infer not only occupation span but also population size, residential movements, contemporaneity between sites, and sociopolitical complexity (Varien and Mills, 1997; Varien and Ortman, 2005; Varien and Potter, 1997).

What has rarely been discussed, however, is the fact that equation 3.2 is effectively a model of temporal mixing, with the parameter t representing the temporal resolution scale of the assemblage formed through discard. Many of the insights gained from accumulation research can thus be applied to the question of resolution and underdetermination.

For instance, accumulation models tell us that temporal mixing is influenced by how mobile people were, as mobility governs occupation span. The Holocene trend toward increased sedentary life, and thus longer occupation span, probably marked a trend toward decreasing resolution of archaeological assemblages.

Accumulation research also describes how spatial mixing increases with population density. For any given amount of time, denser populations will discard more material than sparser ones, because there are more objects in the systemic context in a denser group. For instance, more cooking pots are in use at any given time in a hamlet of eight households than in a hamlet of four households. If the households share the same refuse area, the ceramic assemblage produced by the larger hamlet will have a lower spatial resolution than the one produced by the smaller hamlet. Again, the Holocene trend toward larger settlements, from camps to villages, and from villages to cities, increased the opportunities for spatial mixing and may have resulted in assemblages with lower spatial resolution.

Indeed, the use of secondary refuse amplifies the effect of both occupation span and group size on the resolution of the archaeological record. Most of the archaeological material we find is in *secondary refuse* areas—that is, in places other than where it was used (Schiffer, 1972). *Primary refuse* areas, places where objects are discarded at their location of use, are probably rare and occur only in locations that were occupied for brief periods of time and by small groups, such as the sites where kills were butchered (Schiffer, 1972). When a location is occupied for more than just a few days, the accumulation of discarded items quickly begins to interfere with daily activities and will need to be cleaned up (Schiffer, 1972, 1987). In fact, given the danger they pose, surfaces that were used to chip stone tools were probably frequently and thoroughly cleaned, so that many "activity areas" identified on the basis of lithic debris are probably refuse areas (Schiffer, 1987, 65).

All known societies discard stuff outside their use location. In a survey of discard practices of 79 societies from the Human Relations Area Files, Patricia Murray (1980) found that primary contexts occur only in outdoor locations that are occupied by migratory groups and for less than a season (and even then, some objects are still discarded in secondary refuse areas). In contrast, sedentary and semisedentary groups that occupy a site for at least a season discarded items in refuse areas outside the family living space, where the material accumulates (the only exception are the Chippewa, who either burn their refuse or throw it to their dogs). Murray's work is in line with Schiffer's hypothesis that "with increasing site population (or perhaps site size) and increasing intensity of occupation, there will be a decreasing correspondence between the use and discard locations for all elements used in activities and discarded at a site" (Schiffer, 1972, 162). It also confirms that trash does attract more trash: people tend to dump trash where others have previously dumped trash (Schiffer, 1987, 62), and an initial dumping episode can lead to the development of a dumping area that can be used for a long period of time, further increasing the mixing of artifacts. There are few reasons to think that discard behaviors should have been much different in the past, and that items did not accumulate in secondary refuse areas in most places and time periods.

Again, the Holocene trend toward an increased use of formal dumping areas, culminating in modern sanitary landfills, also furthers the trend toward a decrease in the temporal resolution of archaeological remains. This trend, however, may have been counterbalanced by an increased investment in architecture that spatially segregates secondary deposits. For instance, secondary refuse areas, such as floor layers or middens, are often closely associated in space with individual households (Schiffer, 1987, 80), limiting their spatial resolution to a household level.

Another insight from accumulation research is that the resolution of assemblages is inversely correlated with the use-life of the tools that compose them. The use-life of an object (parameter L in eq. 3.2) is its length of service in the systemic context (Schiffer, 1976, 60). Use-life affects both the size of assemblages and the relative proportions of types of objects in the assemblages, as items with a short use-life accumulate faster than items with a long use-life (Shott, 1989b, 2004). As will be discussed later in this chapter, the mixing of tools with different use-lives can affect dramatically the relative frequencies of types of objects in archaeological assemblages.

However useful accumulation research is in sharpening our intuitions about the impact of discard on resolution, using discard equations to measure the temporal resolution of archaeological assemblages remains a perilous exercise. In principle, equation 3.2 should allow us to use the size of an archaeological assemblage to infer the occupation span it represents, and thus its temporal resolution. But in practice, converting the size of an assemblage into an occupation span is not straightforward, as the tool use-life (L) and average number in systemic context (S) can take a wide range of values.

Many factors influence the use-life of tools, including the function of the object, frequency of use, physical characteristics, reuse, recycling, household inventory size, transfer value, and cultural context (Shott, 1989b, 2004; Varien and Mills, 1997; Varien and Potter, 1997). Similarly, the number of objects in systemic context also depends on many variables, including the object use-life itself (Shott, 1989b, 2004; Varien and Mills, 1997; Varien and Ortman, 2005; Varien and Potter, 1997). For instance, one household may possess more cooking vessels than others because it is wealthier or because one of its members is a potter. In addition, objects may not be replaced as soon as they break. The artifact inventory may be replenished only when raw material or time to manufacture replacements becomes available (M. Nelson, 1991). Given how sensitive the use-life and the number in systemic context are to environmental and cultural contexts, it should not come as a surprise that both variables vary tremendously in the ethnographic record (Shott, 1989b, 2004; Varien and Mills, 1997; Varien and Potter, 1997).

The cross-cultural variation in both L and S is so vast that any estimate of occupation span that is based on the number of objects discarded can be wrong by several orders of magnitude. Mark D. Varien and Barbara J. Mills (1997) have compiled data on ceramic use-life and vessel frequency in systemic context among 19 groups from the Americas and Africa (see also Shott, 1996, for a discussion of the determinants of pottery vessel use-life). Table 3.1 shows the minimum and maximum use-life and vessel frequency for cooking pots and noncooking pots in Varien and Mills's cross-cultural dataset (cooking and noncooking pots are analyzed separately because whether or not a pot is used for cooking affects both its use-life and its frequency in systemic context). Using these values, we can calculate the upper and lower bounds for the occupation duration for an assemblage that contains a certain number of vessels. The rightmost column of table 3.1 shows the number of years needed for an assemblage of 100 vessels to accumulate for different combinations of minimum and maximum use-life and frequency per household observed ethnographically. For instance, 0.57 year (about 6.5 months) is necessary for 100 vessels with a use-life of 0.2 years and a frequency per household of 34.1

TABLE 3.1. Duration of occupation necessary for a household to discard 100 pots, for various values of pot use-life and number of pots per household in systemic context

Function	Use-life (years)	n/household	Duration (years)
Cooking pots	0.2	0.2	100
	0.2	34.1	0.57
	17.8	0.2	8900
	17.8	34.1	52.19
Noncooking pots	0.2	0.1	200
	0.2	24.6	0.81
	20	0.1	2000
	20	24.6	81.3

Note: These values are the minimum and maximum observed in a cross-cultural dataset (Varien and Mills, 1997) for cooking and noncooking pots. For instance, the minimum use-life for a cooking pot in the cross-cultural dataset is 0.2 year, and the minimum number of pots per household is 0.2. Duration of occupation is predicted using equation 3.2.

to be discarded by a household. These calculations show that for any given number of vessels in an archaeological assemblage, the duration of occupation can vary by three orders of magnitude, from 10^{-1} to 10^3 years.

A workaround for these problems is to parametrize the discard model to a particular culture and time period using well-dated archaeological contexts (Varien and Mills, 1997; Varien and Ortman, 2005; Varien and Potter, 1997). For instance, Varien and Mills (1997) used the Pueblo I site Duckfoot, in Mesa Verde, Colorado, to estimate an annual household accumulation rate of cooking-pot sherds. The site of Duckfoot has been completely excavated, and its occupation history is well understood. It was occupied by three large households over a period of 20–25 years, during which the households each discarded between 5323 and 6654 grams of cooking-pot sherds. Varien and Mills used this rate of accumulation as a baseline for accumulation rates of cooking-pot sherds in the Mesa Verde region during the 800s. With this baseline, they estimated occupation spans at other Pueblo I sites. They estimated that four of the sites, small hamlets, were occupied for 15–19 years, 23–29 years, 10–12 years, and 19–24 years and that a fifth one, the large village of Rio Vista, was occupied for 27–34 years, suggesting that the Pueblo I sites of the Mesa Verde region have a temporal resolution that ranges from one to three decades.

In a subsequent paper, Mark Varien and Scott Ortman (2005) used the Duckfoot site accumulation rate to estimate the occupation span of 19 sites,

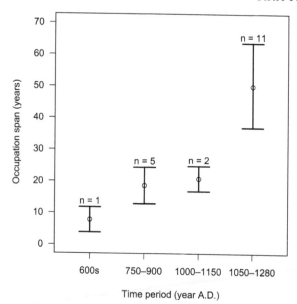

FIGURE 3.1: Average occupation span and confidence intervals for Pueblo sites in Mesa Verde, Colorado, estimated using size of cooking-pot assemblages. These data illustrate how discard models can be used to estimate the temporal resolution of archaeological assemblages. They also show the trend toward longer occupation over time, a feature of the global Holocene archaeological record. (Adapted from Varien and Ortman 2005.)

most of them small farming hamlets. They found that the earliest site in their sample, a residence occupied in the 600s, was occupied for less than 10 years (less than a generation), but that over the next centuries occupation span increased to about 20 years and, by 1100, to about 50 years (fig. 3.1). The spatial resolution scale of the ceramic assemblages could also be measured in terms of the number of households they represent. In the case of these 19 sites, the spatial resolution ranged from one household (in the case of the smallest site) to 16 households (in the case of the largest site). Their result is thus emblematic of the Holocene trend toward greater sedentism and coarser resolutions in the archaeological record.

There are several limitations to applying the method developed by Varien and his colleagues to estimate occupation span and, by the same token, the temporal resolution of archaeological assemblages. The method requires an archaeological case, like the Duckfoot site, that allows us to calibrate the discard equation. It also requires a thorough excavation of sites, since occupation span will be underestimated if objects are left unexcavated. The method also requires that the population size of the site is known. In the case of Pueblo sites, this is done by counting the number of pit structures, which covaries with the number of households. What is more, archaeologists of the American Southwest have fine-grained chronological tools that allow them to distinguish households that were occupied at the same time from households that were occupied sequentially. In most other archaeological contexts, the number of occupants at a site is an unknown quantity, and it is impossible to distinguish a site that has been occupied by a large group for a brief period of time from a site occupied by a small group for a long period of time.

In the end, accumulation research tells us that estimating the resolution of assemblages on the basis of their size alone is difficult. Barring independent lines of evidence for the occupation span of a site, our starting assumption should be that the temporal resolution ranges from hours to months in the case of hunter-gatherer sites, from days to months for pastoralist sites, from 10^{-1} to 10^2 years for small-scale farmer settlements, and from 10^2 to 10^3 years in the case of complex societies.

Reoccupation and Reuse   The reoccupation of a site after its abandonment extends the net period of time over which a location is occupied, thus leading to more objects being discarded and potentially mixed. In equation 3.2, the variable *t* represents the *use duration of an archaeological place*—that is, the aggregate amount of time that a location has been occupied, either continuously or repeatedly (Lightfoot and Jewett, 1984). The distinction is important, because it recognizes the possibility of multiple spatially overlapping occupations that are stratigraphically inseparable (Surovell et al., 2009).

Locations with great ecological or strategic advantages, like caves or spring environments, are more likely to be reoccupied (and, when occupied, are likely to be so for longer periods of time), resulting in long and complex occupation histories (Dibble et al., 2016; Varien and Ortman, 2005). This is why even some of the earliest Paleolithic sites show signs of reoccupation (Semaw, 2000) and why sites all around the world are frequently reoccupied over periods of thousands of years.

But reoccupation does more than merely extend the period of discard: it also affects the nature of the mixed assemblages. Occupation hiatus will generate discontinuities in the representation of time in time-averaged assemblages. The composition of a site occupied multiple times may differ from that of a single-occupation site, even when the duration of use is the same. For instance, Surovell's (2009) discard model predicts that sites that have been occupied continuously will have a higher incidence of local material and more debitage relative to nonlocal tools than sites that have been repeatedly occupied.

Reoccupation also leads to mixing when the remains of different occupations are integrated together (Holdaway and Wandsnider, 2006; Schiffer, 1987; Wandsnider, 2008). For instance, the succeeding occupants of a site may situate their hearths and their middens where previous hearths and middens were constructed, resulting in low-resolution features. Similarly, the very presence of lithics on a surface may attract later individuals to occupy the same location and reuse tools and cores left in place (Dibble et al., 2016).

Geological Depositional Processes

The mixing of archaeological material is as much a function of geological processes as it is of cultural ones. At any point in time, any given portion of the earth's surface is dominated by one of three regimes of geological processes: stability, aggradation, and degradation (Rapp and Hill, 2006; Waters, 1992, 60). Periods of stability are marked by insignificant sedimentation and erosion and sometimes soil development (Waters, 1992, 41–43); periods of aggradation, by the accumulation of sediments on the surface; and periods of degradation, by the removal of sediments and soil through erosion.

If the surface of the earth is a recorder of human activity, then stability, aggradation, and degradation affect how well the recorder works. The quality of the archaeological record in a region, whether its scope, its sampling interval, its resolution, or its dimensionality, will closely follow the timing, the number, the magnitude, the duration, and the areal extent of the periods of stability, deposition, and erosion in that region (Waters, 1992).

Mixing during Periods of Stability   Periods of stability are particularly conducive to mixing. When a surface is stable, cultural materials discarded on a surface accumulate with no vertical separation (fig. 3.2, series A). A stability regime thus increases the opportunities for mixing by discard, reoccupation, and preburial disturbances. The frequency and the duration of the stability periods will determine how likely it is that discrete occupations have mixed on the same surface before burial (Barton and Riel-Salvatore, 2014; Waters, 1992, 97). All other things being equal, archaeological assemblages that are found on the surface of paleosols will have a coarser resolution than assemblages found embedded in sedimentary columns.

Mixing during Periods of Aggradation   During a period of aggradation, mixing occurs when the rate of cultural discard is faster than the rate at which sediments accumulate (Ferring, 1986). When sedimentation rates are slow, cultural remains that represent different moments in time may accumulate

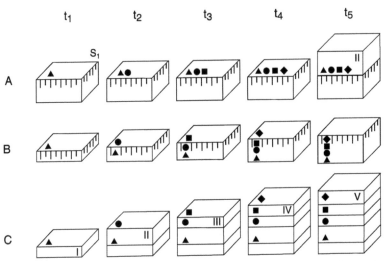

FIGURE 3.2: The deposition of archaeological remains in soils and sediments. Series A shows archaeological remains (the triangle, round, square, and diamond symbols) accumulating on a stable soil surface
S_1 over time t_1 to t_4 and buried by sediments at t_5 (unit II). Series B shows archaeological material being
incorporated into a cumulative soil profile. Sedimentation rate is slow, and soil forms at the same pace as
sediments are deposited. The result is artifacts that are vertically superposed in a layer of soil with little
vertical separation between them. Series C shows artifacts accumulating during a period of aggradation.
Sedimentation rates are fast, and artifacts deposited at different points in time are deposited in different
strata and are clearly separated vertically. (Adapted from Waters, 1992.)

and mix on the same surface or be separated by such a thin layer of sediments
that it will be difficult, or impossible, to separate them into discrete assemblages (Ferring, 1986; Waters, 1992) (fig. 3.2, series B). In contrast, when sedimentation rates are rapid relative to the rate of discard, the remains that are
discarded at different points in time will be separated vertically in the sediment
column, resulting in stratified sites with different cultural horizons (fig. 3.2,
series C). Thus, for any given sedimentation rate, there is a critical waiting
time that has to separate two occupations (or discard events) for them to
form two stratigraphically discrete (i.e., spatially separated) assemblages.
How long this critical waiting time is depends on (1) sedimentation rates,
(2) consistency of sedimentation, and (3) excavation methods.

Sedimentation rates vary, even within the same depositional environment (Sadler, 1981; Schindel, 1980). For example, fluvial sedimentation rates
in North America vary from slow (<0.1 cm/year), to moderate (0.1–0.5 cm/
year), to rapid (0.5–1.0 cm/year), and to very rapid (>1.0 cm/year) (Ferring,
1986).

Consistency of sedimentation refers to the fact that sediments accumulate not consistently but sporadically (Schindel, 1980). For instance, alluvial

sediments may accumulate at a rate of 0.25 cm/year but do so only seasonally. The consistency of sedimentation can be captured by a parameter, C, that specifies the probability that sedimentation occurs at any given time (Schindel, 1980). For instance, when $C = 1$, sediments accumulate continuously (i.e., 100% of the time). Similarly, when $C = 0.5$, sedimentation occurs only half of the time (e.g., six months a year).

The excavation methods deployed at a site (a form of analytical lumping; see below) determine how much vertical separation is needed for archaeologists to recognize stratigraphically distinct assemblages. For example, if an archaeologist excavates using arbitrary layers of 10 cm, she would need a layer of at least 10 cm of sterile matrix in order to detect a discontinuity in the vertical distribution of artifacts.

Together, rates, consistency of sedimentation, and excavation methods specify the critical waiting time that has to separate two events to avoid their mixing. For example, when sediments accumulate at a rate of 0.5 cm/year, when they do so consistently ($C = 1$), and when archaeologists need 10 cm of sterile sediments to recognize distinct cultural levels, the critical waiting time is 20 years: events that are separated by less than 20 years will be mixed stratigraphically.

In most sedimentary contexts, the critical waiting time is on the order of 10^1–10^2 years. Figure 3.3 shows the critical waiting times for rates of sedimentation that vary from slow to rapid (the x-axes), consistency of sedimentation that ranges from $C = 1$ to $C = 0.1$ (the rows), and vertical separations of 1, 5, and 10 cm (the columns). Within this parameter space, the critical waiting times vary from an order of magnitude of 10^0–10^2 years. Critical waiting times in the 10^0-year range (i.e., 1–9 years) demand precise excavation methods that can detect a sterile layer of 1 cm between two cultural levels and, less realistically, highly consistent sedimentation rates ($C = 1$). Most regions of the parameter space are defined by critical waiting times of 10^1 years (i.e., 10–99 years) and are generally greater than 20 years. Longer critical waiting times on the order of 10^2 years (100–999 years) dominate either when a vertical separation of 5 cm is needed and sedimentation is inconsistent ($C = 0.1$) or when vertical separation of 10 cm is necessary and sedimentation occurs half of the time ($C = 0.5$).

Mixing during Periods of Degradation   Erosion can also lead to the mixing of archaeological material. For instance, at the site of Ccurimachay, a rock shelter in Peru, material from the preceramic and the ceramic period that had been deposited over thousands of years became mixed as gravity pulled it downslope from the rock shelter for 20–300 meters (Rick, 1976).

Like gravity, moving agents of erosion can pick up archaeological objects that had been discarded in different places, transport them, and deposit them

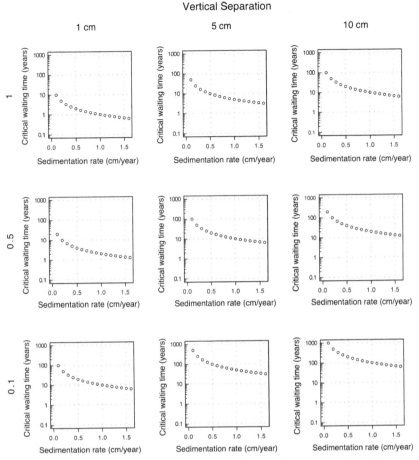

FIGURE 3.3: The effect of sedimentation rate on the temporal mixing of discrete events. The critical wait-ing time (y-axes) is the minimum amount of time that has to separate two events, such as two occupations, in order for them to be separated vertically by at least 1 cm (left column), 5 cm (center column), or 10 cm (right column) of sterile sediments, given a sedimentation rate (the x-axes) and a consistency of sedimenta-tion (rows). The region below the critical line is the region in which mixing occurs, and the region above the line is the region in which the two events will generate assemblages that are vertically discrete.

together in the same location. Whether an object is transported or not by the moving agent of erosion depends on the size of the former and the velocity of the latter. The Hjulström diagram shown in figure 3.4 was developed to de-scribe the effect of water velocity on sedimentary material, but it also applies to archaeological material, since cultural objects can behave as natural sedi-ment particles (Waters, 1992). The diagram shows that the larger an object is, the faster the water flow has to be in order to move it. This means that water flow will differentially affect different types of archaeological remains. Charred

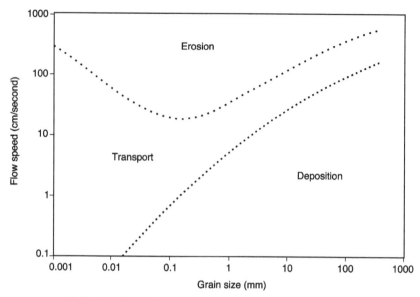

FIGURE 3.4: The Hjulström diagram shows the relationship between water velocity, the size of sedimentary material, erosion, transportation, and deposition. The upper curve is the critical velocity for entrainment—that is, dislodging and starting to move sediment particles. The lower curve is the velocity below which lifted particles settle. The area in between defines the zone at which particles are transported. Transportation velocity can be lower than erosion (entrainment) velocity because entrainment requires more energy than transportation. An entrained particle will continue to move even though flow speed drops below erosion velocity, as long as it does not drop below the critical deposition velocity (Waters, 1992).

seeds and pieces of charcoal can be transported at lower velocity than flakes or hearthstones. Objects transported by moving water are deposited again when water velocity decreases below a certain threshold that also depends on the object size (the lower line in fig. 3.4). It is during this redeposition phase that mixing occurs. In environments such as braided rivers, archaeological material buried in the channel banks can be eroded and transported by the river and deposited in the same sandbar. For instance, archaeologists excavating the site of Double Adobe, in Arizona, found milling stones and handstones along with late Pleistocene faunal remains—mammoths, horses, camels, and bison. Whereas the assemblage had originally been interpreted as a Clovis plant-processing site, a careful geoarchaeological study of the site showed that the megafaunal remains had been eroded from older sediments that made up the banks of a Holocene braided stream but that the artifacts were eroded from the surface of banks and bars, before being deposited together (Waters, 1986a, 1986b).

Similarly, erosion can lead to the spatial mixing of material within an archaeological site, by shuffling and reworking remains that are on or near the surface of the site (Waters, 1992). For instance, in Alaska, a moving glacier reworked the remains left at the site of Hidden Falls, reorganizing the spatial arrangement of hundreds of artifacts and a hearth feature (S. Davis, 1989) and leaving behind an assemblage with a site-wide spatial resolution.

DISTURBANCE PROCESSES

Disturbance processes mix archaeological remains by displacing objects after they have been discarded, either while they sit on the surface of the ground (preburial processes) or after they have been buried (postburial processes). In both cases, the extent to which objects are mixed depends on the object's durability. Durable objects, like ceramics and stone tools, are more susceptible to being mixed by disturbance than nondurable objects such as wood and textiles. Indeed, durable objects can reside longer near the surface, where disturbance processes are most active, and can endure multiple disturbance events.

Preburial Disturbance

Archaeological remains that sit on or near the surface are exposed to all sorts of disturbance processes. Human activity, such as tillage and trampling, disperses artifacts and features and distorts their spatial arrangement (Schiffer, 1987). Similarly, the scavenging of artifacts and the salvaging of material for construction lead to the mixing of material from different occupations. Nonhuman agents like animals, insects, and worms are also known to pick up and displace portable objects (Nash and Petraglia, 1984; Schiffer, 1987), and moving agents of erosion like water and wind can displace artifacts over hundreds of meters (Butzer, 1982; Rick, 1976; Wood and Johnson, 1978). By displacing objects horizontally, these disturbance processes not only modify the spatial distribution of artifacts but also mix together objects that are associated with different activity areas within a site or different occupation events or that are even from different sites.

Postburial Disturbance

Objects that have been buried below the surface are far from safe. Below the surface, a large array of agents may disturb soil and sediments, along with the artifacts that are embedded in them (Butzer, 1982; Schiffer, 1987; Wood and Johnson, 1978). These agents include plowing, burrowing animals and insects, tree fall, freeze-thaw action, frost cracking, and expanding and contracting clay in soil. They can move artifacts up, down, and horizontally and reorient, sort, and even concentrate them into discrete layers and patches (Wood and Johnson, 1978).

As is the case with preburial disturbance, the horizontal displacements generated by postburial disturbance can mix material associated with unrelated events and destroy the spatial patterning of sites (Waters, 1992). And the vertical displacements can mix material from temporally distinct horizons into the same horizon. Over time, different processes that pull objects into different directions can homogenize the distribution of material into the ground. For example, in clay soils in regions with a wet and a dry season, artifacts are pushed upward as the clay matrix swells during the wet season and fall downward through the cracks that open during the dry season (Butzer, 1982; Waters, 1992; Wood and Johnson, 1978). With enough time, such processes can mix together horizons that were originally separated by sterile sediments and generate continuous vertical distributions that are uniform (as in fig. 3.5) or even unimodal (i.e., clustering around the same horizon, as when a tree that is toppled by a storm can pry buried artifacts from the ground and integrate them into the same surface; Butzer, 1982; Waters, 1992; Wood and Johnson, 1978).

What is more, the resulting unimodal vertical distributions can be so narrow as to look like an occupation horizon itself. Archaeologists P. Jeffrey Brantingham, Todd A. Surovell, and Nicole M. Waguespack (2007) have built a series of mathematical models of postburial mixing that allows them to explore the effect of vertical mixing in the cases when (1) artifacts move locally (e.g., from 30 to 31 cm below the surface); (2) artifacts move nonlocally (e.g., from 30 to 15 cm); (3) movement is symmetrical (upward and downward movements are equally likely); and (4) movement is asymmetrical

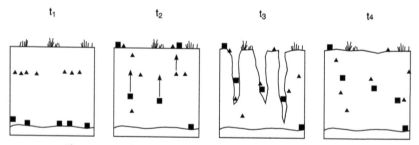

FIGURE 3.5: The movement of artifacts due to the expansion and contraction of clay matrix leads to the mixing of archaeological horizons. Artifacts from two different occupations form two discrete horizons in a clay matrix (t_1). Expansion and contraction of the clay matrix cause the upward movement of artifacts (t_2). Artifacts fall into cracks that form as the clay contracts (t_3) and become sealed again in the clay matrix, forming homogenized cultural debris (t_4). (Adapted from Butzer 1982; Waters 1992.)

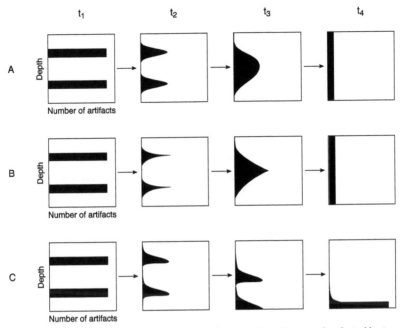

FIGURE 3.6: The major stages in the postdepositional mixing of two discrete archaeological horizons. At time t_1, two discrete horizons are buried in a sedimentary section. (A) Under conditions of symmetrical local mixing, artifacts dissipate vertically in a Gaussian manner across the profile (t_2). Eventually, they form a single unimodal distribution (t_3). At equilibrium, artifacts are distributed uniformly (t_4). (B) Symmetrical nonlocal mixing leads to concave distributions (t_2) and (t_3), which also converge on a uniform distribution at equilibrium (t_4). (C) Local mixing with a bias for downward movement (i.e., asymmetrical) leads to skewed distributions (t_2 and t_3). At equilibrium, the artifacts have accumulated at the base of the section, with a trailing tail above the base of the section (t_4). (Adapted from Brantingham, Surovell, and Waguespack, 2007.)

(such as when upward movement is more likely than downward movement). Brantingham and his colleagues found that the movement of discrete cultural horizons begins with a period of dissipation that is characterized by a normal distribution of artifact depth when movement is local and symmetrical (fig. 3.6A, t_2) and by a concave distribution when the movement is symmetrical but nonlocal (fig. 3.6B, t_2). In the case of asymmetrical movement, the cultural horizon dissipates to form a skewed distribution (fig. 3.6C, t_2). Eventually, the artifacts that belong to different cultural horizons dissipate into one single unimodal distribution (t_3, fig. 3.6). When this happens, the boundaries between the horizons have effectively disappeared and the material associated with them is mixed. With even more time, the distribution of artifacts reaches an equilibrium state marked by a uniform distribution in the case of symmetrical movement (t_4, fig. 3.6A and B) or an accumulation in either the lowest layer in the case of downward-biased movement (t_4, fig. 3.6C) or the highest layer in the case of upward-biased movement. The lesson: even the presence of a vertically narrow horizon is not a guarantee that the assemblage represents a single occupation event.

ANALYTICAL MIXING

Of course, the resolution of the archaeological record is not just a function of cultural and natural site formation processes—it is also a function of the way archaeologists analyze it. How archaeologists construct their analytical units also influences their ability to separate material temporally and spatially, effectively mixing archaeological material above and beyond the mixing generated by depositional and disturbance processes. In fact, by the time archaeologists publish the information that they collected in the field, their data have gone through several rounds of analytical mixing. Some of this mixing will be due to analytical lumping and some of it will be due to imprecision in dating techniques.

Analytical Lumping

Analytical lumping occurs when archaeologists lump together archaeological information in order to create analytical units. The first place where analytical lumping occurs is in the field. The excavation methods used at a site determine what the minimal resolution of archaeological units can be. For instance, sites are often excavated using arbitrary units, such as a a 1 × 1 meter square grid that is excavated by layers of 10 cm. When an excavator bags together the artifacts he found in a 1 × 1 m unit, he creates an analytical unit with a spatial resolution of 1 m². Unless the provenience of the artifacts within the unit is recorded, any spatial patterning that existed within that square meter is lost. Similarly, when artifacts that come from the same 10 cm layer are bagged together, a unit with a temporal resolution equal to the amount of time represented in that 10 cm layer of sediment has been created. Likewise, the excavation and the lumping of artifacts on the basis of geological layers can easily generate analytical units that represent hundreds, if not thousands, of years (Binford, 1982; Dibble et al., 2016; Stern, 1994, 2008).

A second round of analytical lumping takes place inside the archaeologist's computer. When archaeologists analyze and publish their data, they lump together the units that were created in the field to create new units that are more relevant to their research questions. For instance, they may lump together the artifacts that belong to the same cultural level, activity area, or the same site. In doing so, they create units with coarser spatial and temporal resolution.

And these horizon-, activity-, or site-level units may themselves be lumped into larger multisite units. These larger units may be constructed to compare different time periods, regions, or ecological habitats, for example. The zooarchaeologist Lee Lyman (2003) uses a fictional case to illustrate the different ways in which these multisite units can be created. Imagine that there are six archaeological assemblages that come from six different sites. The sites were deposited sequentially over time and span two cultural time periods. The sites are also geographically close to each other but are located in different ecological habitats. One way to analyze these data is to plot each site against its age, with different symbols denoting the different sites. This preserves the temporal and the spatial distinctiveness of the assemblages and does not lump them. But other ways of analyzing the same data leads to analytical lumping. First, the data may be plotted against age, but using the same symbol for all the assemblages, thereby muting the spatial distinctiveness and resulting in spatial lumping. Second, the data may be plotted with different symbols to maintain their geographical distinctiveness, but according to the cultural period to which they belong; this is temporal lumping. Third, the data may be averaged according to the cultural time period and the summary data plotted against the average age of the lumped assemblages, resulting in both spatial and temporal lumping. All these different ways of lumping archaeological data are found in the extant archaeological literature (Lyman, 2003).

Lumping by cultural time period is especially prevalent in archaeology. Archaeologists have long divided human history into chronological cultural units such as "periods," "phases," "stages," "horizons," or "cultures" (Willey and Sabloff, 1993). These cultural time periods are often the main unit of analysis in archaeological publications: it is these cultural time periods that archaeologists compare when they keep track of prehistoric settlement patterns, subsistence, social organization, mortuary behavior, or trade and exchange patterns, with the events that took place within each time period treated as contemporaneous.

Typically, these cultural time periods are on the order of 10^2–10^4 years long and, spatially, upward of 10^3 km^2. For instance, the *Outline of Archaeological Traditions*, a worldwide database of archaeological data assembled by archaeologist Peter Peregrine, divides world prehistory into 88 "archaeological traditions" that he defines as "a group of populations sharing similar subsistence practices, technology, and forms of socio-political organization, which are spatially contiguous over a relatively large area and which endure temporally for a relatively long period" (2001, iv). These 88 traditions roughly reflect the way archaeologists divide the prehistory of the regions in which they work. The temporal resolution of these archaeological traditions spans four orders

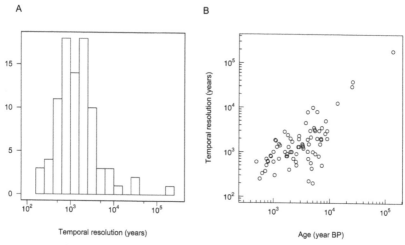

F I G U R E 3.7: The temporal resolution of the analytical units in the *Outline of Archaeological Traditions* (Peregrine, 2001). (A) Frequency distribution of the temporal resolution of the units. (B) The temporal resolution of the units as a function of their age. The age of a unit is the midpoint of the unit's age range.

of magnitude, ranging from 200 to 185,000 years, with a median of 1400 years (fig. 3.7A). What is more, resolution increases with age: the traditions of the last millennium are 10^2 years long, whereas traditions that are tens of thousands of years old have resolutions on the order of 10^4 years (fig. 3.7B) (see also Eighmy and LaBelle, 1996). Spatially, the archaeological traditions have resolutions that range from small regions, like the "Jomon" tradition, which is confined to the islands of Japan; to large regions, such as the "South Indian Neolithic," which spans the Indian subcontinent south of the Ganges River valley; and to continents, such as the "Early Paleo-Indian" tradition, which spans North America and parts of Mesoamerica and South America.

Analytical lumping is useful and here to stay. First, some research questions themselves demand some level of analytical lumping. For instance, assessing the diet of a population demands that we lump and average together many contexts representing single meals (Lyman, 2003). Second, analytical lumping gives us statistical power. Pooling assemblages into groups allows archaeologists to create samples that are sufficiently large to conduct statistical analyses. Third, cultural time periods are an important dating tool. Very often, the only way to date an archaeological context is to assign it to a cultural time period based on the *fossil directeurs* it contains (M. E. Smith, 1992).

The same issues of statistical power and dating explain, in part, why cultural time periods become longer the further we go back in time. Because of preservation biases, younger sites are more frequent than older sites (Surovell and Brantingham, 2007; Surovell et al., 2009), which means that archaeologists working on recent time periods do not need to cast as wide a net to gain statistical power as the archaeologists working on older time periods do. In addition, a greater diversity of material is preserved in younger sites, making it easier to identify and recover types that are associated with a specific time period. Finally, more chronometric dating techniques, and more precise ones, are available for the more recent time periods, allowing us to divide the recent past into finer temporal units.

Imprecision of Dating Techniques

Analytical lumping also arises from the imprecision of the dating techniques used by archaeologists. Every dating technique comes with a certain precision—a level of unsystematic errors that is due to measurement errors and to uncertainty in the calibration process. The precision of a dating technique specifies the shortest amount of time over which the technique can distinguish between contemporary and noncontemporary events. In the case of the dating techniques used by archaeologists, this precision ranges from 10^0 to 10^4 years.

The most precise dating technique used by archaeologists is dendrochronology. With a precision of 10^0 years, dendrochronology can distinguish between two events that are 1 year apart from each other. But however precise it is, dendrochronology is available in only a handful of regions in the world and can be applied only to the Holocene period. Other dating techniques are much less precise because they are prone to multiple sources of unsystematic errors. For instance, the precision of obsidian hydration dates varies significantly because of errors in the measurement of hydration thickness of the obsidian samples, errors in the estimate of hydration rate, and errors in the temperature history of the sample (Pierce and Irving, 2000), such that late Holocene samples from Rapa Nui have been dated with a precision of 30 years (1σ; $2\sigma = 60$ years) (Stevenson, Ladefoged, and Novak, 2013), whereas the dates for an Early Jomon occupation in Japan have a longer error of 178 years (1σ; $2\sigma = 356$ years) (Nakazawa, 2016).

And some dating techniques become less precise with time. Thermoluminescence and optically stimulated luminescence have a 1σ precision of 5%–20% of the mean age estimate (Forman, 2000). A 5% precision means that as early as 400 BP, events that are separated by 20 years are seen as geologically contemporaneous. By 2000 BP, the 5% precision is enough to make events separated by 100 years appear synchronous. And by 20,000 BP, a 5% precision translates into an error of 1000 years.

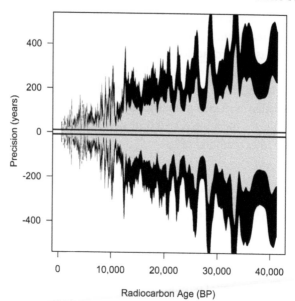

FIGURE 3.8: Precision of calibrated radiocarbon dates based on the IntCal13 calibration curve (Reimer et al., 2013) with Oxcal (Bronk, 2009). Estimates are based on a sequence of radiocarbon dates sampled from 500 BP to 40,000 BP with a 50-year interval and a precision of ±0.5%. The gray area represents the calibrated range of one standard deviation (68%), and the black area represents the calibrated range of two standard deviations (95%). The two horizontal black lines represent a range of one human generation (±10 yr).

Argon-argon (^{40}Ar/^{39}Ar) dating has similar precision. Argon-argon dating plays an important role in paleoanthropology and Paleolithic archaeology, although its time range of applicability extends well into the Holocene (Renne, 2000). With best practices, Ar-Ar dating can produce age estimates that are as precise as 0.1% (2σ) (Erwin, 2006). This translates into precisions of 20 years for a 2000-year-old sample (in theory; in practice the precision of such young dates is likely to be larger than 0.1%), of 100 years for a 10,000-year-old sample, and of 10,000 years for a 1-million-year-old specimen.

Last but not least, radiocarbon dating, the most widely used archaeological dating technique, has a precision that varies from 10^2 to 10^3 years. The error of a radiocarbon date has two components: (1) the uncertainty in the measurement of the radiocarbon content of the sample and (2) the uncertainty in the calibration curve (Trumbore, 2000). Using accelerator mass spectrometry to measure radiocarbon content allows precisions that range from 0.5% to 2% (Trumbore, 2000). Figure 3.8 shows the precision of calibrated radiocarbon dates assuming radiocarbon ages with the smallest error possible, ±0.5%, and using the IntCal13 calibration curve (Reimer et al., 2013). The figure shows that over the entire time range of applicability of radiocarbon dating, the precision is greater than 20 years 99.5% of the time. The median precision over the last 10,000 years is ±88 years (1σ); and beyond 10,000 BP, ±320 years (1σ).

The Effects of Mixing of Archaeological Data

The primary effect of mixing, whether it is caused by depositional processes, disturbances processes, or analytical lumping, is to reduce the resolution of archaeological units. It reduces temporal resolution (i.e., time averaging) by lumping material associated with activities that took place at different points in time. And it reduces spatial resolution (i.e., space averaging) by mixing material associated with activities that took place at different points in space.

The impact of mixing on archaeological data has been little studied. It is not that archaeologists have failed to recognize the time- and space-averaged nature of the archaeological record. For more than 30 years, archaeologists have been using the metaphor of "palimpsest" to refer to sites where successive activities have been superimposed and reworked (Bailey, 1981; Binford, 1981; Dibble et al., 2016; Foley, 1981), a metaphor that is still in use today (e.g., Malinksy-Buller, Hovers, and Marder, 2011). Entire theoretical approaches, such as time perspectivism (Bailey, 1981, 1983, 1987, 2007, 2008; Davies, Holdaway, and Fanning, 2016; Dibble et al., 2016; Fletcher, 1992; Holdaway and Wandsnider, 2008; T. Murray, 1999; T. Murray and Walker, 1988; Stern, 1993, 1994; Wandsnider, 2008), are premised on the time-averaged nature of the archaeological record. Time averaging has also been discussed, albeit obliquely, in the research on occupation span and assemblage composition (e.g., Ammerman and Feldman, 1974; Schiffer, 1974, 1987; Shott, 1989b, 2004, 2008) and on sample size and assemblage richness (e.g., Grayson, 1984; Jones, Grayson, and Beck, 1983; Kintigh, 1989; Meltzer, Leonard, and Stratton, 1992; Rhode, 1988; Shott, 1989a, 2008, 2010). And there are a few studies that have examined the effect of time averaging on specific domains of archaeological research, such as foraging theory (Lyman, 2003), lithic *chaîne opératoire* (Vaquero, 2008), and cultural transmission (Garvey, 2018; Porčić, 2015; Premo, 2014). But unlike paleontologists (e.g., Behrensmeyer, Kidwell, and Gastaldo, 2000; Fürsich and Aberhan, 1990; Hunt, 2004; Kidwell and Behrensmeyer, 1993; Kidwell and Flessa, 1996; Kidwell and Holland, 2002; Kowalewski, 1996; Kowalewski and Bambach, 2003; Kowalewski, Goodfriend, and Flessa, 1998; Olszewski, 1999; Sadler, 1981; Schindel, 1980; Wilson, 1988), archaeologists have not developed a thorough and general theory of how mixing affects their data and their capacity to test hypotheses.

ORDINARY AND SIGNIFICANT MIXING

A good theory of mixing first needs to recognize that the extent to which mixing interferes with archaeological inferences depends on the phenomenon of interest. The resolution of an archaeological unit can never be zero—every archaeological assemblage is mixed at some scale or another, be it minutes or centimeters. Thus, the question is not so much whether a unit is mixed or not but whether it pools together contexts that are, given the phenomenon of interest, related or unrelated. The pooling of related contexts constitutes "ordinary" mixing, and the pooling of unrelated contexts is "significant" mixing (Kowalewski, 1996; Kowalewski and Bambach, 2003).

By and large, ordinary mixing is desirable. For every phenomenon, there is a Goldilocks level of mixing—an amount of mixing that is just right to avoid sampling errors while avoiding mixing with unrelated contexts. Paleontologists call this level of mixing the "minimum duration of time averaging": the time period over which samples must be pooled in order for the composition of a fossil assemblage to reflect that of a living community (Kidwell and Bosence, 1991; Martin, 1999). The same reasoning applies to archaeological contexts. For example, as mentioned previously, an archaeologist needs to pool together a certain number of individual meals in order to reconstruct the diet of ancient people (Lyman, 2003).

By contrast, significant mixing is undesirable because it distorts our view of a given phenomenon. Imagine a historian characterizing the diet of seventeenth-century American yeomen using a database of recipes that included, unbeknownst to her, the recipes of twentieth-century American cook Julia Child—she would conclude that *homards thermidors* and *éclairs aux chocolats* were standard fare on the colonies' tables.

A good theory of mixing also acknowledges that it is not always possible to distinguish ordinary from significant mixing. As we saw earlier, measuring precisely the amount of mixing due to discard is impossible without accurate estimates of discard rates and population size. Similarly, the geological markers that indicate number and duration of occupations, such as hearth area, artifact/feature ratios, and the thickness of anthropogenic soils (for more examples, see Malinksy-Buller, Hovers, and Marder, 2011; Wandsnider, 2008), or that signal pre- and postburial disturbance are not always present and, when they are, may go undetected by archaeologists. Even an excavation technique like *décapage*, whereby sediments are removed one thin layer at a time, can be misleading, as the surface it exposes is arbitrary and may still contain material that has been deposited at different points in time (Dibble et al., 2016). The same is true of lithic refits, which tell us the order in which flakes have been removed from a core but not the duration of time over which the reduction took place: multiple knappers may very well have reused the same block of stone at different points in time and for different purposes (Dibble et al., 2016).

In a seminal volume on natural formation processes published in 1987, David T. Nash and Michael D. Petraglia pointed out that archaeologists often operate under the belief that the effects of natural processes can be readily identified and separated from the cultural ones. Thirty years later, not much has changed. Although awareness of natural formation processes is greater today than it was in 1987, their effect is often treated as nonexistent, negligible, or easily identifiable. Still today, archaeologists often assume that sets of contemporary traces can be separated from palimpsests and interpret nonrandom spatial patterns strictly in cultural terms. And still today, the prevailing assumption among archaeologists is that archaeological contexts can be treated as ordinarily mixed unless proven otherwise. For instance, the archaeologist who classifies sites and site areas on the basis of their "function" is assuming that the site is not significantly mixed (Holdaway and Wandsnider, 2006; Shott, 2008, 2010). Similarly, aberrant radiocarbon dates are often labeled as "outliers" and discarded. But these dates are outliers only if one assumes that the dated assemblage has a fine-grained resolution. Once that assumption is removed, the outlying dates become important pieces of information about the formation of the site and its resolution (Seymour, 2010).

THE REPRESENTATION OF TIME IN MIXED ASSEMBLAGES

The effect of temporal mixing on archaeological data depends not only on the duration of mixing but also on the representation of time in the mixed unit. For instance, a temporally mixed unit is not necessarily an averaged representation of the period of time it represents (Stern, 2008). Instead, the representation of time in the unit is likely to be uneven and skewed (Kowalewski and Bambach, 2003; Martin, 1999).

Several factors conspire to create a nonuniform representation of time in time-averaged assemblages. Because of changes in discard rate, some moments will be overrepresented in a mixed assemblage and others will be underrepresented. Similarly, sequences of occupation and abandonment will create discontinuities in the temporal coverage of a mixed unit. Taphonomic biases can also create a discontinuous and skewed representation of time. Whole sections of time may have disappeared because of a degradation event. More subtly, age-biased taphonomic destruction, by which older material is more likely to

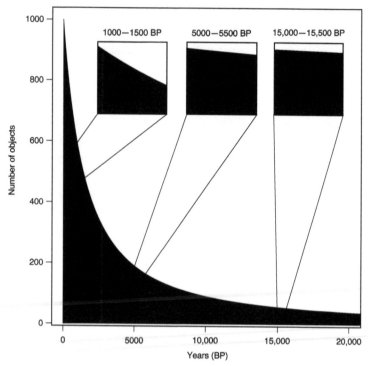

FIGURE 3.9: Taphonomic loss leads to internally skewed mixed assemblages that are biased toward younger specimens. Assume that every year, 1000 artifacts are discarded at a site. Objects are lost to taphonomic destruction at a rate of $\rho = 1.3925309/(2176.4+t)$ (Surovell et al., 2009), where t is the age of the objects. The inserts show the internal temporal structure of assemblages with a resolution of 500 years sampled at different time periods.

have been destroyed than younger material, will create mixed assemblages in which older material is underrepresented and younger material is overabundant (Olszewski, 1999). For instance, the global process of loss of preservation identified by Surovell et al. 2009) generates an exponential-like loss of material over time (see chapter 4). This process can generate mixed units that are skewed toward younger material, even when the input rate of artifacts remained constant over time. Imagine that 1000 artifacts are discarded at a site every year between 20,000 BP to the present. Without taphonomic loss, the temporal frequency distribution of artifacts would be uniform: the artifacts dated to, say, 15,000 BP would be as frequent as those of any other year. But if artifacts are destroyed at the rate identified in Surovell et al. 2009, the temporal frequency distribution of artifacts will be heavily skewed toward younger time periods (fig. 3.9). The mixing of material deposited at different points in time, and after taphonomic destruction, will result in a skewed representation of time inside mixed units. For instance, a unit that lumps the material deposited between 1500 and 1000 BP will contain 18% fewer objects from the year 1500 than from the year 1000 (fig. 3.9). Thus, the average value of an archaeological trait in this temporally mixed unit will be closer to what the value was in the year 1000 BP than it was at 1500 BP. Since the rate of taphonomic loss decreases with age, the representation of time in mixed units becomes more uniform as we go back in time. In a unit that lumps the material deposited between 5000 and 5500 BP, the artifacts from 5500 BP are 9% less frequent than those from 5000 BP material (fig. 3.9). Similarly, the unit mixed over 15,000 and 15,500 BP is only slightly skewed, with a difference between the frequency of oldest and youngest material of less than 4%. This suggests that the effect of taphonomic loss on the internal structure of mixed assemblages decreases with the increasing age of the assemblage and will affect more severely the younger parts of the archaeological record.

MIXING AFFECTS THE SIZE AND THE COMPOSITION OF ANALYTICAL UNITS

The most immediate effect of mixing on archaeological data is to increase the number of specimens in archaeological units. This is why archaeologists often use artifact density to monitor occupation span, occupation intensity, or sedimentation rates—all agents of mixing.

The inflation of frequencies caused by mixing can affect archaeological interpretations. There are objects and features whose mere presence is thought to be a meaningful anthropological signal about the past, such as objects made of exotic material, ritual paraphernalia, or specialist tools, and that are used as smoking guns to confirm hypotheses. Mixing, by increasing the frequency of these objects from zero to one or from one to many, can affect the way an archaeological context will be interpreted.

The effect of mixing on frequencies can also make different depositional sequences look similar archaeologically. Table 3.2 illustrates how three faunal assemblages representing different subsistence strategies can look equivalent archaeologically (de Lange, 2008). In the first scenario, adult and juvenile prey are acquired at a constant rate over the entire occupational history of the site. In the second scenario, the population shifts gradually from a focus on adult prey to a preference for younger and older animals. In the third scenario, there is a punctuated shift from the exclusive deposition of adults to the exclusive deposition of young and old animals, with a 150-year hiatus in between. When mixed, the three deposition sequences yield assemblages with the exact same frequency of adults and young/old animals. If the three mixed

TABLE 3.2. Three scenarios of animal procurement history

Time (years)	Scenario 1	Scenario 2	Scenario 3
0–50	10 adults, 3 young/old	15 adults	10 adults
50–100	10 adults, 3 young/old	22 adults, 1 young/old	30 adults
100–150	10 adults, 3 young/old	16 adults, 1 young/old	40 adults
150–200	10 adults, 3 young/old	11 adults	20 adults
200–250	10 adults, 3 young/old	21 adults, 3 young/old	—
250–300	10 adults, 3 young/old	6 adults, 4 young/old	—
300–350	10 adults, 3 young/old	4 adults, 6 young/old	—
350–400	10 adults, 3 young/old	3 adults, 5 young/old	8 young/old
400–450	10 adults, 3 young/old	2 adults, 6 young/old	15 young/old
450–500	10 adults, 3 young/old	4 young/old	7 young/old
Mixed content	100 adults, 30 young/old	100 adults, 30 young/old	100 adults, 30 young/old

Source: de Lange 2008.

Note: Because of mixing, the assemblages formed under these three scenarios contain the same frequency of adult and young/old animals, even though they represent different animal procurement strategies.

assemblages represented three different cultural levels at a site, an archaeologist may be tempted to interpret the sequence as indicating a stable foraging strategy over a long period of time, even if in reality there were significant short-term adaptive changes in procurement strategies (de Lange, 2008).

In addition to inflating frequencies, mixing also affects the composition of analytical units. For instance, mixing can distort the composition of a ceramic assemblage by changing the relative frequency of decorative styles that are present in it, by inflating the diversity of tempers observed, or by increasing the variance in vessel size. These three classes of effects are discussed in more detail below. But it is worth nothing here that the extent to which mixing affects the composition of an archaeological unit depends on how much the parameter of interest changed over the period of mixing (or the area of mixing in the case of space averaging). For instance, if a community manufactured ceramic pots in the exact same way over a period of 500 years, the variance in pot thickness will remain the same whether an assemblage is time averaged over 50 or 500 years. But if the ceramic tradition evolved over the course of the site occupation, then variance in pot thickness will increase with time averaging. Effectively, time averaging and space averaging act as nets that capture temporal and spatial variation in material culture and incorporate this variation into the same analytical unit. The longer the period of mixing is or the larger the area of spatial mixing, the larger the net is, and the more likely it is to capture some variation in the parameter of interest.

Mixing Skews Relative Frequencies

Relative frequencies, whether of tool types, ceramic styles, or body parts in a faunal assemblage, occupy a central place in archaeological analyses. Unlike absolute frequencies, relative frequencies are more robust to sample size issues and, because of that, are assumed to capture better what happened in the past. But for relative frequencies to capture past dynamic contexts with fidelity, the level of mixing has to be just right—not too little mixing but also not too much.

A factor like tool use-life can keep the level of mixing on the "too little" side of the scale: that is, not enough mixing to qualify as "ordinary mixing" (see above). Relative frequencies change with occupation span, not only because longer occupations capture a wider range of activities, but also because they capture the discard of objects with long use-lives (Shott, 2004). In fact, tool use-life alone can make relative frequencies shift significantly during an occupation, even when the activities performed at a site remain constant.

Archaeologist Michael Shott (1989b) examined the joint effect of occupation span and tool use-life on the composition of tool assemblages left at !Kung San camps. He found that the occupation span of !Kung San camps is shorter than the typical tool use-life: the use-life of !Kung San tools is counted in hundreds and thousands of days, whereas their camps are occupied for a few dozens of days at the most (Shott, 1989b, 13, table 3). As a result, he found no agreement between occupation span, the size of assemblages, and their composition (table 3.3). Neither did he find an association between the activities conducted at the camps and the composition of assemblages, since most tools used at a camp were not discarded there. For example, there is no correlation between the number of days the !Kung San hunted or the number of kills they made and the number and types of stone tools left behind. Shott's results are at odds with the prevailing working assumption—especially among archaeologists studying foragers—that the size and the composition of an assemblage reflect, somehow, the occupation span as well as the activities that were conducted at the site (Shott, 1989b). In the case of the !Kung San camps, both tool assemblage size and the frequency of the different tool types underdetermine the occupation span of the camp and the set of activities that took place there.

Thus, relative frequencies will change from the moment a site is occupied to the moment when the tools with the longest use-life are discarded, even if the activities performed at the site remain constant. Take a simple scenario

TABLE 3.3. Occupation span, activities, and tool assemblage composition from !Kung San camps

Occupation span				Activities		Tool assemblage		
Camp	Days	Man-days	Person-days	N hunts	N kills	Size	Types	Hunt
1	8	13	29	6	9	21	5	1
2	9	18	36	9	2	0	0	0
3	11	28	54	28	8	15	1	0
4	20	48	98	—	—	13	3	0
5	2	4	8	2	4	0	0	0
6	3	9	15	9	2	0	0	0
7	10	40	75	27	9	5	2	1
8	30	180	330	—	—	3	2	2
9	2	6	10	5	6	4	1	0
10	12	84	156	—	2	14	2	2
11	3	21	39	—	—	11	2	1
12	3	15	30	11	3	8	1	0
13	5	25	50	16	1	31	2	0
14	7	56	115	—	6	18	4	2
15	1	5	10	2	2	3	1	0
16	6	36	72	17	4	7	2	0

Source: Adapted from Shott, 1989B; original data from Yellen 1977.

Note: Occupation span is measured in terms of the number of days a camp is occupied, the number of man-days, and the number of adult-person-days. The hunting activities are measured in terms of the number of man-days in which hunting occurred and the number of animals obtained. The composition of the tool assemblage is captured by the total number of tools involved in collecting and hunting, the number of types of tools left, and the number of tools used for hunting.

imagined by Shott (2008): an individual performs, every day, the same set of activities. This set of daily activities involves five tools that each belong to a different class of tools. The tools from the five classes have different use-lives. Those from the first class have a use-life of one day, the tools from the second class have a use-life of two days, the tools from the third class have a use-life of three days, and so on. Within the same site, the relative frequencies of tool class will change every day as tools that are at the end of their use-life are discarded. For example, from the first to the fifth day of occupation, the relative frequency of class 1 tools decreases from 100% to 50% (table 3.4). Relative frequencies will continue to change until a site has been occupied long enough for the tools with the longest use-life to be discarded, at which point relative abundances remain stable over time. If the individual were to occupy different locations for different amounts of time, the assemblages she would leave behind would have different compositions, even though she performed the exact same activities at each site. Thus, variation between archaeological contexts in the relative frequency of artifacts is driven not necessarily just by variation in behavior but also by the joint effect of occupation span and tool use-life (Shott, 2008; Surovell, 2009). In this example, the relative frequencies of the five tool classes settle on the fifth day, after which there is "enough" mixing to capture the composition of the forager's toolkit, so that the relative frequencies after 100 days of occupation are not much different from what they were after 5 days.

Forces such as disturbance processes and analytical lumping are particularly conducive to the problem of "too much mixing." For instance, the analytical-lumping schemes used by zooarchaeologists can easily affect prey-abundance ratios in ways that can change archaeological interpretations (Lyman, 2003). As long as the types of objects discarded remain the same over time, and the frequency of these types remains constant, temporal mixing will not affect relative frequencies (fig. 3.10A). But if there is variation in the type of object discarded or in the frequency at which they are discarded, then the relative abundances in the mixed assemblage will be unlike those at any given point in time (fig. 3.10B). The time-averaged assemblage will likely overestimate the importance of objects that have been discarded at a constant rate over time while underestimating the importance of those whose frequency has fluctuated over time. For instance, in figure 3.10B, type D dominates the mixed assemblage even though it was never the most important type of object. Conversely, type F is relatively rare in the mixed assemblage even though it was initially the dominant type. Finally, even a single short-term fluctuation in abundance ratios is sufficient to make the relative frequencies of the mixed

TABLE 3.4. A model of the joint effect of occupation span and tool class use-life on toolkit composition

Occupation span	N (relative frequency)					Richness
	Class 1	Class 2	Class 3	Class 4	Class 5	
1 day	1 (100%)	—	—	—	—	1
2 days	2 (66%)	1 (33%)	—	—	—	2
3 days	3 (60%)	1 (20%)	1 (20%)	—	—	3
4 days	4 (50%)	2 (25%)	1 (15%)	1 (15%)	—	4
5 days	5 (50%)	2 (20%)	1 (10%)	1 (10%)	1 (10%)	5
100 days	100 (44%)	50 (22%)	30 (13%)	20 (8%)	10 (4%)	5

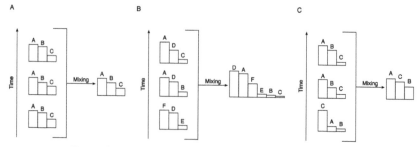

FIGURE 3.10: Temporal mixing affects relative frequencies in the archaeological record. Mixing affects the relative frequencies of types of objects (denoted by uppercase letters). (A) Relative frequencies remain constant over the period of mixing. (B) Relative frequencies fluctuate over time. (C) Relative frequencies are generally constant but with a short fluctuation. (Adapted from Fürsich and Aberhan, 1990.)

assemblages unlike any of those that prevailed at any given point in time (fig. 3.10C) (Fürsich and Aberhan, 1990).

And of course, the impact of mixing on relative frequencies undermines not only the interpretations of these relative frequencies but also the other methods that depend on them, such as the use of power-law frequency distribution of cultural traits to identify modes of social learning (Porčić, 2015; Premo, 2014) and frequency seriation (de Barros, 1982).

Mixing Affects Richness

Figure 3.10B also illustrates how mixing influences richness—that is, the number of types or classes in an analytical unit (e.g., Bobrowsky and Ball, 1989; Kintigh, 1989; Shott, 2004). Archaeologists use richness in things such as raw-material types, prey items, and style to infer all sorts of parameters about past populations, such as occupation span, level of mobility, group size, extent of social networks, subsistence patterns, and modes of social learning.

Richness is underestimated when there is not enough mixing. Perhaps the excavated portion of a site is not representative of the unexcavated portion. After all, types of objects are not homogeneously distributed within a site, so that richness may increase as the excavated area is expanded (e.g., Grayson, 1984; Jones, Grayson, and Beck, 1983; Kintigh, 1989; Meltzer, Leonard, and Stratton, 1992; Rhode, 1988; Shott, 1989a). Or maybe there is not enough mixing because a site has not been occupied long enough to incorporate low-probability activities (Grayson and Delpech, 1998; Yellen, 1977) like the butchering of a prey item that is rarely captured or the acquisition of a rare metal. Similarly, the same joint effect of occupation span and tool use-life that affects relative abundance also affects richness (Ammerman and Feldman,

1974; Schiffer, 1975, 1987; Shott, 1989a, 1989b, 2008), leading to the so-called "Clarke effect"—that is, the tendency for richness to increase with a settlement's occupation span (Schiffer, 1987, 54–55).

Conversely, richness may be inflated by mixing. Measures of richness are particularly sensitive to inflation by mixing because an object has to appear only once in a context to contribute to its richness. Thus, even a very short-term fluctuation in past behavior can be enough to increase richness. Imagine a group that returns to a location every fall. During the first 99 years of the site's history, the same three species are butchered at the site: bighorn sheep, pronghorn antelope, and mule deer. In the 100th year, however, the group kills a deer, a jackrabbit, and a marmot. Because of this once-in-a-century fluctuation in the group's foraging, the mixed assemblage left at the site has a prey taxa richness of 6—twice the richness that prevailed for 99% of the time the site was occupied (Grayson and Delpech, 1998).

Richness inflation can create deceiving patterns. An assemblage that contains six different tool types may have been left by one group that stuck to the same toolkit over the occupation history of the site (fig. 3.11, left), but it may also have been left by multiple groups that drew from different small toolkits (fig. 3.11, right). Interpreting richness is thus challenging. The archaeologist Manuel Vaquero (2008) used the site of Abric Romaní in Spain to examine how the analytical lumping of lithic artifacts by stratigraphic units influences archaeological interpretations. Archaeological remains found embedded in one of the stratigraphic levels of the site exhibited five clusters. Vaquero shows that lumping the lithic remains from these clusters on the basis of their presence in the same stratigraphic unit, as archaeologists often do, would lead to wrong conclusions about the activities that took place at the site. For instance, since the mixed assemblage would contain the by-products of every step of a *chaîne opératoire*—cortical products, flakes of all sizes, debris, cores, and retouched flakes—one may conclude that the entire reduction sequence was carried out at the site. But this conclusion could be wrong. Cores and tools can be brought to a site at any stage of a *chaîne opératoire*. It is possible that objects representing different states of reduction were introduced into the different clusters. Indeed, cores are present in only two of the five clusters, and another cluster lacks both cores and retouched tools. The impression that the entire reduction sequence was conducted at the site would be an artifact of the mixing of spatially distinct contexts. In turn, second-order inferences about settlement pattern, site type, or occupation span that are contingent on the presence of a complete *chaîne opératoire* would also be false. Because of how sensitive richness is to mixing, it is not surprising that the statistical

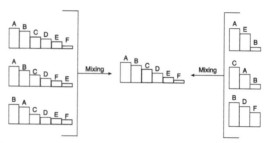

FIGURE 3.11: Mixing can increase richness in equifinal ways. The mixing of contexts produced by a stable group (left) and the mixing of contexts produced by different groups (right) generate an equally diverse assemblage. (Adapted from Fürsich and Aberhan, 1990.)

methods that rely on richness perform poorly with time-averaged assemblages (e.g., Premo, 2014).

Mixing Increases Variance

Mixing inflates variance in continuous traits such as morphometric variables (Bush et al., 2002; Hunt, 2004; Lynch, 1990; Wilson, 1988). It does so by collapsing the variance that exists *within* contexts (e.g., within a population at time t) into the variance that exists *between* contexts (e.g., between a population at time t and $t + 1$). The greater the between-contexts variance is, the more inflated the variance in a mixed assemblage will be.

The magnitude of the between-contexts variance is a function how (1) the mean and (2) the variance of a trait have changed between contexts. Imagine the mixing of five archaeological contexts that were deposited at different points in time. In one scenario, the mean of a trait represented in these five contexts (e.g., hearth circumference) changed linearly and gradually over time, while the variance around the mean remained constant. If these five contexts were to be mixed, the variance in the time-averaged assemblage would be much larger than it ever was at any single point in time: the distribution of the trait would be wider, flatter, and converging to a uniform distribution (fig. 3.12A). The same goes for a scenario in which the mean of the trait fluctuates over time (fig. 3.12B). A punctuated shift in the mean or a gap in the sequence of mixed contexts would lead to a bimodal distribution (fig. 3.12C and D). And a stable mean but changing variance would result in a time-averaged distribution in which the central values are overrepresented (fig. 3.12E). Finally, the same processes can also occur in space (fig. 3.12F). What is more, none of the mixed distributions shown in figure 3.12 have a shape that betrays their time-averaged nature—even the bimodal distributions in panels

C and D could be interpreted as signaling the coexistence of two classes of objects with overlapping morphologies.

We can actually go a step further than the qualitative model shown in figure 3.12 and develop a formal model that we can use to test archaeological hypotheses. Imagine a continuous archaeological trait that is normally distributed (e.g., vessel size). As in figure 3.12A and B, the mean of the trait changes over time but its variance remains stable. Appendix A shows that the

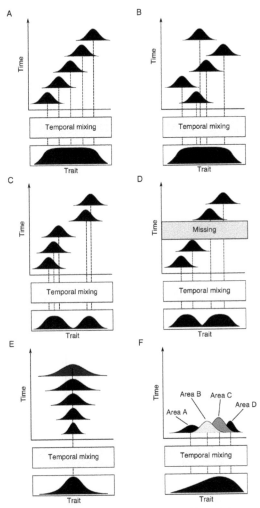

FIGURE 3.12: Mixing increases variance in archaeological assemblages. (A) Linear change in mean of the distribution of a trait. (B) Fluctuating mean. (C) Slow change with punctuated period of rapid change. (D) Noncontinuous time averaging. (E) Stable mean and fluctuating variance. (F) Variances in space. (Adapted from Bush et al., 2002; Wilson, 1988.)

variance inflation in mixed assemblages is not so much a function of how much the mean of a trait changed between each context but rather of the overall dispersion of the mean over the period of mixing. A trait that oscillates rapidly but within a narrow range of values will not inflate the variance of a mixed assemblage to the same extent as a trait that evolves more slowly but over a larger range (Hunt, 2004).

The next step in building the model is to describe how the mean of a trait changes over time. There are many ways to do this, but the safest bet is to assume that the distribution of the archaeological trait is temporally autocorrelated: the distribution of the trait at time t depends on its distribution at the preceding time step, $t - 1$. Such autocorrelation can arise for multiple reasons, including social transmission, stylistic drift, or environmental constraints. And let us further assume that the mean of the trait distribution shifts every time step by an incremental amount drawn from a random distribution with a mean μ_{step} and a variance δ_{step}. Thus, μ_{step} represents the directionality of change. When $\mu_{step} = 0$, the change in the mean is an unbiased random walk. When $\mu_{step} < 0$, the mean of the trait decreases over time, and when $\mu_{step} > 0$, it gets larger. Similarly, δ_{step} represents the pace of change—how volatile change is around the trend set by μ. When $\delta_{step} = 0$, change occurs at a constant pace set by μ_{step}, and when it is greater than zero, change over time becomes a random walk. The expected inflation of the variance due to mixing is thus

$$(3.3) \qquad E\left[V_M\right] = \frac{t^2 \mu_{step}^2}{12} + \frac{t \delta_{step}^2}{6}.$$

This equation (see Hunt, 2004, for derivation) partitions the variance inflation that is due to μ_{step} (the first term on the right side of the equation) and to δ_{step} (the second term). The variance inflation due to μ_{step} increases exponentially with time, whereas the inflation due to δ_{step} increases linearly with it (fig. 3.13). Thus, all other things being equal, the directionality of change (i.e., the magnitude of μ_{step}, whether positive or negative) is a more potent driver of variance inflation than unbiased changes (i.e., the magnitude of δ_{step}). In an archaeological context, this means that the problem of variance inflation is particularly acute during the periods in which a trait is evolving in a systematic manner (e.g., projectile points become smaller). In contrast, traits that do not evolve in one particular direction, either because they are neutral or because they are bounded by functional constraints, are more robust against the effect of mixing.

We can use this model of variance inflation to answer all sorts of questions about the effect of mixing on variance. For instance, how fast does the statistical signal of a punctuated event decay in a mixed assemblage?

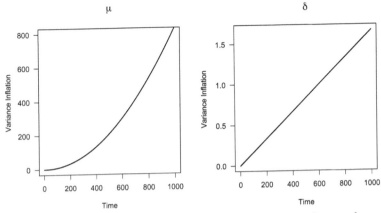

F I G U R E 3.13: Variance inflation in a time-averaged assemblage due to changes in the mean of a population through time due to the effect of μ_{step} (left-side plot and first term on the left side of eq. A4) and δ_{step}
(right-side plot and second term on the left side of eq. A4). Both μ_{step} and δ_{step} are set to 0.1.

Imagine that two groups split from the same population and settle in two
different locations, sites A and B. Initially, because they come from the same
cultural group, the pottery produced at sites A and B is the same. However,
the second group abandons its new home after just 1 year, whereas the first
group remains at site A for decades. Assuming that the ceramic produced every year is mixed with the ceramic discarded the previous years, and that the
ceramic tradition evolves over time, how many years must pass before the ceramic assemblages left at sites A and B become statistically different? In other
words, how long will it take for the distribution of a trait at a particular point
in time (the first year of occupation of sites A and B) to become swamped by
the distributions produced at other points in time? This is an important question because an archaeologist may use, for example, the similarity between
two assemblages, or a lack thereof, to infer whether the occupants of sites A
and B are culturally related or not.

The answer to this question depends on how the ceramic tradition changed
over time, or, in terms of the Markov model above, the values of μ_{step} and δ_{step}.
We can simulate the Markov model to explore how the time to signal loss var-
ies with μ_{step} and δ_{step}. Let us assume that the ceramic vessel volume during the
first year of occupation of the two sites was normally distributed, with a mean
$m = 1500$ mm and a standard deviation $\sigma = 50$ mm. The vessel volume evolves
over time: the mean volume size m changes every year by an increment γ that
is drawn from a normal distribution with mean μ_{step} and variance δ_{step}. Every
year (and every time step in the simulation), 50 pots are produced and discarded in a midden where they mix with the vessels discarded in the previous

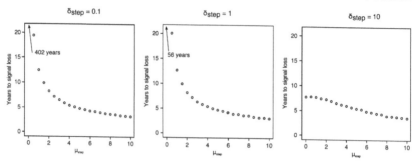

FIGURE 3.14: Simulation of the effect of mixing on the loss of short-term signal due to variance inflation. The short-term signal is the distribution of a trait at the beginning (year 1) of the simulation (see
text for details). Y-axes are truncated: the values under arrows indicate years to signal loss when $\mu = 0$.

years. At the end of each year, the simulation tests whether the mixed assemblage is significantly different from the distribution produced in the first year
at both sites A and B—that is, a normal distribution with $m = 1500$ mm and $\sigma = 50$ mm—using a two-tailed Kolmogorov-Smirnov test. The simulation tallies
the time to signal loss: the average number of years until the probability that
the two distributions have been drawn from the same distribution is less than
5%. The simulation is repeated until the sample error of time to signal loss is
smaller than 1% of the mean time to signal loss.

The δ_{step} parameter is set to 0.1, 1, or 10, and the values of μ_{step} explored range
from 0 to 10. Note that these values correspond to smaller rates of change than
the rates observed empirically in the archaeological record. The typical ratio
between the initial and final value of a cultural trait after 1 year in the archaeological record of North America is 1.022 (Perreault, 2012), which, for a trait with
a mean of 1500 mm, would correspond to an increment of change of 33 mm. In
other words, the values explored in the simulation are conservative.

The results show that even when assuming that archaeological traits
evolve more slowly than they do in reality, short-term statistical signals are
lost after just a few years of mixing (fig. 3.14). The signal of the distribution
of vessel volume during the first year is most robust when traits change following an unbiased random walk and when variance in step size is very small
(i.e., $\mu_{step} = 0$). This is because unbiased change is marked by a series of increase and decrease that cancel each other out, so that the distribution of
the trait evolves slowly. But when variance in step size is increased, mixing
can bury the statistical signal within decades ($\delta_{step} = 1$) or even within years
($\delta_{step} = 10$). Traits that change following a biased random walk $\mu_{step} > 0$ will be
greatly affected by mixing, and time to signal loss is for the most part shorter
than 10 years, independently of step variance δ_{step}.

These results do not bode well for one of the most common exercises in lithic and ceramic studies, the statistical comparison of continuous traits between sites. Indeed, continuous traits are so vulnerable to mixing that even a small difference in occupation span (or mixing by any other way) may be enough to lead archaeologists to overstate the behavioral significance of the difference between the sites.

Mixing Confounds Associations and Correlations

Mixing also creates associations and correlations that never existed in past dynamic contexts. For example, the discard of tools over a long period of time can generate new correlations between tool types. What is more, these new correlations may very well resemble those generated by behavioral patterns (Ammerman and Feldman, 1974). For instance, gravity-induced disturbances at the Ccurimachay rock shelter in Peru created statistically significant associations between zones of high bone density and projectile points that look like they are the result of human behavior (Rick, 1976). These spurious associations and correlations are particularly likely to affect durable objects, as they are more susceptible to mixing by disturbance.

The effect of mixing on associations is not necessarily an adverse one. Some correlations are visible only in assemblages that have been sufficiently mixed. Unless two traits are perfectly correlated, there is always some random error in a regression model: data points do not all fall perfectly along the regression line. These random errors can dominate and mask correlations, but they can be muted by mixing. In figure 3.15A, the correlation between two variables is not visible when the fine-grained contexts C_1, C_2, C_3, and C_4 are looked at individually but is revealed by mixing the four contexts together.

Alas, mixing can also have the opposite effect: it can mask existing correlations under random noise. A correlation that exists within a context can disappear when mixed with other contexts in which the correlation is absent or is in the opposite direction. For example, in figure 3.15B the correlation that exists in context C_1 disappears when mixed with contexts C_2 to C_4. Correlation may be present in some contexts and not others because of differences in social, cultural, or environmental contexts. For instance, a correlation between access to metal tools and wealth may fade away as metal becomes cheap and abundant. Finally, mixing can mask correlations when the intercept of the models varies between contexts (fig. 3.15C) or when the strength of the correlations changes between contexts (i.e., the slope; fig. 3.15D).

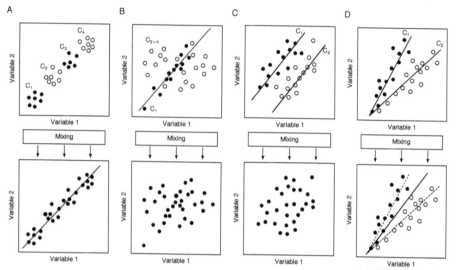

FIGURE 3.15: Mixing affects correlations between variables. (A) Mixing can unravel correlation. In fine-grained contexts (C_1, C_2, C_3, C_4), the correlation between variable 1 and variable 2 is dominated by random errors, but it becomes visible in a time-averaged assemblage. (B) The correlation between variables 1 and 2 in context C_1 disappears when mixed with contexts C_2 to C_4, in which the variables are not correlated. (C) Mixing conceals correlations when the intercept varies between contexts C_1 and C_2. (D) Mixing affects the slope of the regression model because the slope varies between contexts C_1 and C_2. (Adapted and modified from Bush et al., 2002.)

Mixing Reduces Rates of Change

Mixing also affects the perceived rates of change in the archaeological record. Mixing can make change appear to have been rapid whereas it was gradual, and vice versa. The pace of change in the archaeological record, and whether it was gradual or abrupt, has a lot of bearing on how archaeologists interpret the past. In some cases, such as the extinction of megafauna at the end of the Pleistocene, the rate of change—whether it was gradual or rapid—is considered to be a smoking gun in and of itself.

Mixing can create the appearance of abrupt and systemic cultural change, especially when due to analytical lumping. Different cultural traits that disappeared at different points in time may look like they disappeared at the same time when they are aggregated into the same cultural time period (i.e., the opposite of the Signor-Lipps effect discussed in chapter 4). Indeed, the lumping of archaeological contexts by cultural time period leads to a stepwise pattern of cultural change, according to which change occurs between periods and not within (Frankel, 1988; Lucas, 2005; F. Plog, 1974; Shott, 2015).

Conversely, mixing reduces observed rates of change in continuous traits such as length, thickness, or volume. This inverse correlation between temporal mixing and rates happens when mixing causes the two analytical units from which a rate is calculated to converge toward the same mean trait value, as they can do when traits drift randomly or are under selection (Perreault, 2018). The effect of mixing on rates can be seen in the archaeological record. Rates of change calculated from technological traditions found in the Holocene North American archaeological record, such as Anasazi pit structure depth, Chesapeake pipe stem diameters, and Missouri ceramic vessel wall thickness, are inversely correlated with the duration of the cultural time periods the units are assigned to (Perreault, 2018). For instance, a rate calculated from two units representing two cultural time periods of 50 years will be faster, on average, than rates calculated from two time periods of 500 years (19%–68% slower according to the statistical model fitted to the empirical data) (Perreault, 2018). One of the dominant features of the global archaeological record is that the pace of change appears to decrease as we go back in time. The results presented here suggest that time averaging may be contributing to this pattern, along with the effect of time intervals (Perreault, 2012), as archaeologists lump archaeological material into increasingly longer cultural time periods when they analyze older deposits.

Conclusion

The taxonomy of the forces that shape archaeological data presented in this chapter is, at first glance, unusual. It brings together disparate phenomena that are not usually thought of as belonging together, such as site reoccupation, disturbance by burrowing animals, and dating imprecision. Yet all these phenomena have the same net effect on archaeological data, that of decreasing their resolution. By destroying existing patterns and creating new ones, the forces of mixing influence every aspect of the archaeological record from which archaeologists draw inferences, including the size of assemblages, relative frequencies, richness, variance, correlations, and rates of change.

4

The Forces That Shape the Quality of the Archaeological Record, II: The Loss of Archaeological Data

Most of the information about the human past is missing from archaeological data, because of either preservation loss or observational loss. Preservation loss happens when the remains of the past have not been preserved or have been damaged to such an extent that the information-bearing traces they contain have been obliterated. Observational loss occurs when the physical remains are preserved in the archaeological record but have not yet been discovered or recognized by archaeologists. Both forms of loss have many different causes. Expanding on George Cowgill's (1970) idea of three sampling populations, Michael Collins (1975) identified a series of sampling biases that affect archaeological data:

1. Not all behavior results in patterned material culture
2. Of those that do, not all can enter the archaeological record
3. Of those that do, not all will enter the archaeological record
4. Of those that do, not all will be preserved
5. Of those that do, not all survive indefinitely
6. Of those that do, not all will be exposed by archaeologists
7. Of those that do, not all will be identified or recognized by archaeologists

Biases 1–5 are those that lead to preservation loss, and biases 6–7 lead to observational loss.

Many of the causes and effects of preservation and observational loss on archaeological data and interpretations are well understood by archaeologists. Zooarchaeologists, for instance, have produced over the years a rich body of literature on the preservation and modification of bones (Marean and Spencer, 1991; Grayson, 1984; Lyman, 1994; Reitz and Wing, 2008). Instead of wasting time reinventing the wheel, I dwell in this chapter on those aspects of loss that have not been investigated as thoroughly by archaeologists.

The Causes of Loss

MOST THINGS ARE NEVER RECORDED IN THE FIRST PLACE

Most things from the human past have not left any traces in the archaeological record, for the simple reason that they did not involve material culture. Most behavior, most cultural traditions, most social norms, most historical events, and most psychological, social, demographic, and cultural processes never make it past Collins's third bias and are never recorded archaeologically. This is the single most important factor that explains the gulf that separates the ethnographic record from the archaeological record.

Archaeologists pride themselves on finding clever ways to recover these intangible aspects of the past by identifying how they may correlate, somehow, with material culture. The classic example is "ceramic sociology" (S. Plog, 1978), which sought to infer residence patterns and other aspects of prehistoric social systems from the spatial/temporal distribution of ceramic styles. But to use these purported material proxies to infer what is missing from the archaeological record is to accept standing on shaky scientific ground. Each one of these proxies is a hypothesis in need of verification and, in practice, is never more than tenuously verified. Some proxies are based on ethnographic analogies and thus suffer from the same limitations as analogical reasoning (chapter 1). Others are supported only by an unverified line of reasoning, because to verify them would require an independent line of evidence, which, if it existed, would defeat the need for a proxy in the first place. For instance, we may have good reasons to expect that cultural assimilation should covary with the cessation of imports from the homeland, and nothing is stopping us from interpreting a decline in imports as evidence for assimilation. But there is no way to verify that both variables were indeed causally linked in the past. What is more, even if the proxy was valid, it is, in all probability, an imperfect and noisy one. The reasonings that underlie these archaeological proxies all come with an "all other things being equal" clause—a string of factors that need to be controlled in order for the proxy to be accurate. The cessation of imports from the homeland may indeed covary with cultural assimilation, but it also probably covaries with many other variables. Before an archaeologist can use cessation of imports to infer cultural assimilation, he would need to show that these other covariates cannot explain the decline in imports observed in a region. The problem, however, is that these other covariates may not have left unambiguous traces in the archaeological record either.

Even behaviors that are deeply anchored in material culture have facets that are never recorded archaeologically. Take the production and the use of stone tools. Even though stone tools preserve well archaeologically, we do not understand at even the most basic level how they were used. Studies have repeatedly shown that the form of stone implements does not encode enough information—about the intentions of the knapper, about the actions that contributed to their production, about the number of individuals who contributed to their production, or about whether the form recovered was the one intended as the end product—to allow archaeologists to ascertain what implements were used for, let alone if they were used at all (as is the case with unretouched flakes) (Dibble et al., 2016). The presence of use-wear and residue can mitigate these problems, but not all activities leave use-wear or residues, and the same tool may have had multiple uses during its lifetime (Dibble et al., 2016).

CULTURAL PRACTICES

Cultural practices in the past lead to the loss of information by (1) dissociating remains, (2) destroying them, and (3) affecting their archaeological visibility.

Residential and logistic mobility, reuse, and scavenging can dissociate remains that would have been associated with each other otherwise. When foragers leave a camp and take their tools with them, they are, effectively, dissociating these tools from the remains left behind at the site. This is why some sites contain evidence of flint knapping but no finished products, whereas others contain finished tools but no flakes (Dibble et al., 2016; Schick, 1987b; Turq et al., 2013). Similarly, reuse and scavenging (Schiffer, 1987) disarticulate archaeological contexts and lead to the loss of associations between remains.

Cultural practices also destroy objects and traces. The use-wear left on a tool may be obliterated after the tool is reused. Wood used at a site may be scavenged and burned during a subsequent occupation. Likewise, the way animals are cooked and their carcasses disposed of affects their preservation (Reitz and Wing, 2008). Of course, cultural practices can also have the opposite effect and improve preservation. For instance, the practice of burying the dead shields bodies from surface disturbance processes and improves the chance that they are recovered by archaeologists.

Finally, cultural practices affect the visibility of archaeological remains, thereby influencing the probability that they are observationally lost. The density of artifacts discarded at a site, the area of a site, and whether or not architectural features are present affect the likelihood of their discovery through pedestrian survey or shovel testing (Schiffer and Wells, 1978). Construction of structures, roads, or terraces can also improve archaeological visibility. For instance, in the Aegean, archaeological visibility is heavily determined by the particular terrace construction technique used at a site (Frederick and Krahtopoulou, 2000). Similarly, human remains buried in cemeteries are more likely to be discovered by archaeologists than those that are not.

DETERIORATION

Of the few aspects of the human past that were lucky enough to enter the archaeological record and escape loss by cultural practices, many will deteriorate and disappear well before archaeologists have the chance to record them. Whether or not a material trace deteriorates and the rate at which it does depend on its environment and its intrinsic properties (Schiffer, 1987).

Some environments are more conducive than others to deterioration. For instance, organic material will deteriorate more rapidly in a tropical forest than in a desert (Schiffer, 1987). But the microenvironment that immediately surrounds the remains is as important as the regional environment (Schiffer, 1987). Microenvironments can create, within a regional environment that is normally conducive to preservation, circumstances that are favorable to deterioration. Conversely, they can create opportunities for good preservation even in environments that do not normally facilitate preservation (Schiffer, 1987).

Within any given microenvironment there exists a multitude of agents of deterioration—chemical agents, physical agents, and biological agents that operate on different types of material (Greathouse, Fleer, and Wessel, 1954; Rapp and Hill, 2006; Reitz and Wing, 2008; Schiffer, 1987; St. George et al., 1954). Chemical agents deteriorate archaeological remains by triggering chemical reactions. For instance, oxygen and water in the atmosphere corrode metals, acid soils degrade bones, while basic soils degrade pollen. Physical agents such as moving water and wind can break down, abrade, and dissolve artifacts. Others, like earthquakes, landslides, and volcanoes, can not only damage and collapse architectural features but also favor their preservation by quickly burying them. Physical agents can also move and dissociate objects much as the cultural practices of scavenging and reuse do. For example, flowing water winnows artifacts according to their size and weight, displacing small objects while leaving the heavier ones in place. In a series of experiments, Kathy Schick (1987a) found that flowing water alone could create core-rich deposits in the vicinity of the original site and debitage-rich deposits downstream, with a significant spatial gap in between. Biological agents are the main causes of the decay of organic matter. Living organisms such as fungi, bacteria, and insects destroy artifacts made of wood and plant fibers. Other biological agents do not obliterate cultural remains but modify them substantially. For instance, the gnawing, swallowing, and trampling of bones by animals can remove diagnostic marks on their surfaces and even leave traces that resemble intentional fragmentation and butchering marks.

These agents of decay operate at varying rates depending on the micro-environment and the type of material. While all types of material can deteriorate, including stone and ceramic, they do so at varying rates because they are affected differentially by the different agents of degradation (Greathouse, Fleer, and Wessel, 1954; Rapp and Hill, 2006; Schiffer, 1987).

Degradation also leads to an indirect form of observational loss by destroying datable material. For instance, surface lithic scatters often remain undated because they do not contain organic material that can be dated with radiocarbon. An archaeologist may exclude these undated surface scatters from her dataset, either because there is too much uncertainty about the age of the assemblage or because the analysis she is conducting demands that some archaeological variables be plotted against time. In either case, the information preserved in the scatters, within the context of that study, is effectively lost.

SEDIMENTATION AND SURFACE COVER

Sedimentation rates modulate both preservation and observational loss. Sedimentation leads to preservation loss by regulating rates of deterioration (Ferring, 1986; Waters, 1986a). Several agents of deterioration operate at or near the surface, like sunlight, wind, bacteria, and animals. As archaeological remains are gradually covered by sediments, they become increasingly shielded from these agents of deterioration (Schiffer, 1987, 150–52). Cultural remains are thus more likely to be preserved when discarded during a period of aggradation. Conversely, the material deposited during a period of stability or slow sedimentation will be exposed to a wider range of agents of deterioration (as well as cultural practices such as reuse and scavenging) and for a longer period of time. Objects that are fragile, easily transported, or reusable will be particularly affected by increased residence time on the surface.

Sedimentation generates observational loss by affecting the visibility and accessibility of archaeological deposits. Archaeological sites that are located in areas where sediments accumulate frequently or at a fast rate, such as alluvial plains, are, for all practical purposes, invisible to archaeologists. Whereas remote-sensing technologies do increase the visibility of these sites, they are costly and time-consuming to use. What is more, deposits that are buried deep are difficult to access, as excavating them is costlier than excavating surface sites (Schiffer and Wells, 1978).

Because of erosion, the volume and area of sediments available for archaeologists to excavate decrease with the age of the sediments: that is, older geological deposits are, all other things being equal, rarer than younger ones (Raup, 1979), which, in turn, means that older archaeological deposits are fewer, less visible, and more degraded than younger ones. Sedimentation also gives rise to secular trends in the loss of archaeological material. Since aggradation, stability, and degradation depend in part on climate, climatic fluctuations will generate systematic biases in preservation loss and visibility.

Surface cover also affects archaeological visibility. For example, sites located under a dense forest cover will be difficult to detect by pedestrian survey (Schiffer, 1987). Changes in the levels of oceans and lakes have also made the traces of human activities along coastlines difficult to find. Likewise, the induration of the surface influences site visibility. Artifacts can sink and disappear in loose surface, especially when they are trampled over, so that artifacts are more visible on hard, scoured surfaces than on softer and sandier ones (Wandsnider, 1987).

FIELD METHODS

No matter how careful archaeologists are, information is always lost during the excavation process. Depending on the field methods, some information may be missed, left out, or destroyed. There was a time when screening was not a standard practice, and the excavations conducted decades ago failed to recover traces that, today, are systematically recovered, such as lithic debris and seeds. Likewise, there are surely traces that we do not collect today but that will be systematically recovered in the future. Above and beyond what the current excavation standards are, remains are also selectively recovered on the basis of what the excavator finds important, as well as the costs of transportation, analysis, and curation (Reitz and Wing, 2008).

Leaving portions of sites unexcavated so that future archaeologists can go back and recover what we missed only partially mitigates this problem. After all, the portion of a site left untouched may not be representative of the excavated one. What is more, leaving parts of a site unexcavated is in itself a type of observational loss, albeit a temporary one (Meadow, 1981).

The Effects of Loss

LOSS DECREASES ASSEMBLAGE SIZE AND FREQUENCY OF SITES

In contrast to mixing, the primary effect of preservation and observational loss is to decrease the size of archaeological assemblages. For instance, of the 60 postholes left at a site, only 4 may have been preserved and observed by archaeologists. By the same token, loss also decreases the frequency of archaeological sites.

The decrease in the number of archaeological remains caused by loss has a dramatic effect on archaeological research. The archaeological record is a finite resource, and preservation loss chips it away and irremediably. It is always possible that too many traces have been destroyed for archaeologists to compile samples that are representative of the past. For instance, there may not be enough Archaic period hearths in a region, whether discovered or undiscovered, to estimate accurately what the typical hearth diameter was during that time period. By increasing the rate of sampling errors, loss allows chance to play a disproportionate role in the patterns that archaeologists observe in the record. Because of loss, two assemblages can appear different when they should look the same, or they can appear the same when they should look different. How many of these false patterns created by loss have archaeologists imbued with anthropological meanings?

One false pattern that has been wrongly interpreted by archaeologists is the temporal frequency distribution of radiocarbon dates. Around the globe, frequency distributions of radiocarbon dates from archaeological contexts all show the same peculiar pattern: an exponential increase over time. This pattern is often taken by archaeologists as signaling population growth. But Todd Surovell and his colleagues suggested that the pattern may be instead the result of preservation loss (Surovell and Brantingham, 2007; Surovell et al., 2009). Testing this hypothesis, however, is difficult because the effect of demographic change and the effect of preservation loss on the frequency of radiocarbon dates are equifinal. To circumvent this problem, Surovell et al. looked at the temporal frequency distribution of volcanic eruptions. Volcanism, an abiotic process, is unaffected by demographic or cultural factors. And just as important, there are two independent records of volcanic activity: a record of radiocarbon-dated eruptions in terrestrial sediments and the GISP2 (Greenland Ice Sheet Project) ice core in Greenland. What Surovell et al. realized was that the terrestrial record is subjected to preservation loss, whereas the ice core record is not, and that by comparing the two, we can estimate the rate of preservation loss in terrestrial sediments.

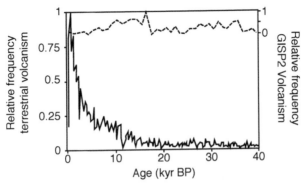

FIGURE 4.1: Temporal frequency distribution of radiocarbon-dated terrestrial volcanic deposits (solid line, data from Bryson, Bryson, and Ruter, 2006) and of volcanic eruptions in the GISP2 ice core from Greenland (dashed line, data from Zielinksi et al., 1996). (Redrawn from Surovell et al. 2009.)

The terrestrial and the ice-core record paint two very different pictures of volcanic activity over time (fig. 4.1). The terrestrial record shows volcanic activity increasing exponentially over time, much like archaeological radiocarbon dates do, whereas according to the ice-core record, volcanic activity has remained fairly constant over the last 40,000 years.

The curve of terrestrial volcanic events can be used as a proxy for the temporal distribution of "geologic opportunities for archaeological sites to exist" (Surovell et al., 2009, 209), that is, for preservation loss. Surovell et al. found that the model that best explains the terrestrial volcanic temporal frequency distribution is one in which the rate of taphonomic loss ρ varies with site age:

$$(4.1) \qquad \rho = \frac{1.3925309}{2176.4 + t},$$

where t is the site age, ρ represents the probability that a site is lost in any given year after its creation. Equation 4.1 says that a site has a 0.06% chance of being lost during its first year (i.e., a 99.94% chance of survival). Following that first year, the annual probability of a site being destroyed becomes smaller and smaller every year. For example, if the site has survived its first 10,000 years, the probability that it is lost during the year 10,001 is 0.01%. This decline in rates of preservation loss with site age makes sense. Since many agents of disturbance and deterioration operate primarily at or near the surface (Surovell et al., 2009), archaeological remains are at greatest risk immediately after they have been deposited. Subsequently, their chance of survival gradually improves as they are blanketed by an increasingly thick layer of sediments.

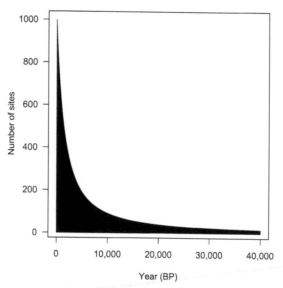

FIGURE 4.2: Taphonomic loss leads to a frequency distribution of archaeological sites that decreases through time. The figure shows the frequency distribution of archaeological sites, assuming that each year, from 40,000 BP to the present, 1000 sites are created and that every year, sites are destroyed with probability $\rho = 1.3925309/(2176.4 + t)$ (Surovell et al., 2009), where t is the site age (see text for details). Of the 40,000,000 sites generated, 3,812,753 (9.5%) have survived to the present, most of which come from the recent past.

The probability of site destruction, ρ, is always very small, but over long time scales its effect adds up: the probability that a site survives not just one but thousands of years is very small. The result is a frequency distribution that increases over time in a curvilinear fashion (fig. 4.2). Surovell et al.'s study suggests that the exponential increase in the frequency distribution of archaeological radiocarbon dates around the globe is primarily due to taphonomic loss.

Surovell et al.'s study also provides us with a useful number: an empirical estimate of ρ, the rate of preservation loss in terrestrial sediments. In the rest of this chapter, I use their fitted ρ value to parametrize models of loss. Of course, this estimate is not without limits. The taphonomic rate ρ is probably not representative of the rates of preservation loss in all types of archaeological contexts, especially those that can be dated by methods other than radiocarbon techniques. What is more, rates of preservation loss will vary microregionally, regionally, and temporally. There may also be secular trends in human prehistory that influence taphonomic loss. For instance, the transition to agriculture was accompanied by permanent architecture and by the production of durable technologies such as pottery, which may have led to a decrease in preservation loss that is not captured by radiocarbon databases. But despite these limitations, Surovell et al.'s estimate of ρ is the best model available to describe how preservation loss affects the archaeological record.

LOSS AFFECTS THE COMPOSITION OF ASSEMBLAGES

The effect of loss on the composition of assemblages is widely recognized and needs to be only mentioned here. Loss introduces sampling errors and systematic biases that, in many ways, have the opposite effect to mixing.

Because of loss, the archaeological record is a highly biased record of the human past. With smaller sample sizes, low-frequency remains, like those associated with tools with a long use-life or behaviors that are rarely performed, will be systematically underrepresented. This will affect the integrity of assemblages (i.e., the extent to which an assemblage represents the totality of the activities that took place; Dibble et al., 2016). Preservation loss will also skew relative-abundance curves toward the most durable objects. To enter the archaeological record and be discovered by archaeologists is a rare event, and objects that have even slightly better chances of being preserved or of being discovered will be dramatically overrepresented in archaeological datasets (Raup, 1979). This, in turn, may lead archaeologists to overestimate the importance of the activities that involve durable material and to underestimate, if not ignore completely, the importance of activities that involve nondurable material. Similarly, loss can differentially affect objects from the same class of material on the basis of their size. For instance, small bones are less likely to survive or to be observed than large bones. In turn, correlations and association between classes of objects that are differentially affected by loss—such as stone tools and plant seeds, large and small body parts, or those objects that people transport as they move in the landscape and those that they leave behind (Dibble et al., 2016)—can disappear completely. And of course, the task of comparing the composition and the content of assemblages will result in underdetermination if the assemblages have suffered from different amounts of preservation loss (e.g., the sites were exposed to different microenvironments) or observational loss (e.g., one site has been excavated more thoroughly than the other).

LOSS INCREASES SAMPLING INTERVAL

Preservation loss and observational loss increase the sampling interval of archaeological data. This is an important fact not only because sampling interval is a major source of underdetermination but also because long hiatuses in archaeological sequences are often interpreted as signaling the abandonment of a region by prehistoric people, with little regard to the possibility that the hiatuses are a statistical artifact resulting from loss (Rhode et al., 2014).

How exactly loss gives rise to the intervals observed in the archaeological record is more complicated than it seems. Gaps are a function of three factors: (1) sample size (e.g., the number of archaeological sites in a database), (2) the spatial and temporal dimensions of the sampling universe, and (3) how much material has been preserved in the archaeological record. These three factors interact in complicated ways, and while archaeologists have some intuitions about why there are gaps in the archaeological record, they lack a mechanistic theory of how gaps arise. In this section I lay the foundation of such theory. I begin by looking at the three factors individually before considering how they interact with each other. For the sake of convenience, I assume that the unit of interest is the archaeological site, but the exact nature of the analytical unit is not important—the term "site" here can refer to any dated and geolocalized archaeological context.

Sample Size Is Inversely Correlated with Time Gaps

Even if the preservation of past cultural remains was total and perfect, archaeologists would still have to contend with gaps in their data. Archaeologists always work from limited samples that are drawn from the larger set of archaeological sites that are available to be discovered, known and unknown—what Cowgill (1970) calls the "physical finds population." This case of observational loss is as much a determinant of the number and the duration of gaps in archaeological datasets as preservation loss is.

The effect of site sample size on time gaps is easy to describe mathematically. Archaeologist David Rhode and his colleagues (2014) developed a model that describes the probability of observing long time gaps in regional sequences of radiocarbon dates. While their goal was to distinguish gaps resulting from sample size from gaps marking true hiatuses in human occupation, their model can be adapted to examine how the distribution of time gaps in regional sequences varies with sample size. The model, which they call the "uniform-frequency model," assumes that all sites have the same probability of being discovered by archaeologists. This assumption allows us to isolate the effect of sample size from covariates like preservation biases, variation in archaeological visibility of sites, and changes in the intensity of human occupation over time.

One counterintuitive result of the model is that despite the fact that sites are uniformly distributed in time, the frequency distribution of time gaps is

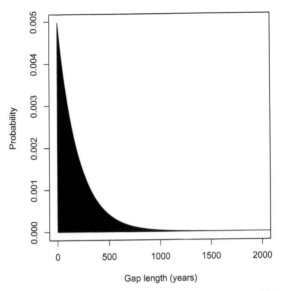

FIGURE 4.3: Probability distribution of gap length in the uniform-frequency model (eq. 4.2) assuming that $\lambda = 0.005$.

not uniform. The uniform-frequency model assumes that the age of archaeological sites is uniformly distributed (hence its name). For instance, a site occupied 600 years ago has the same probability of being sampled as a site that dates back to 5000 years ago. This probability of being sampled is captured by the parameter λ, which represents the average number of sites per year. When $\lambda = 0.01$, the probability of sampling a site of any particular age is 1%. The parameter λ can be estimated by calculating the average number of sites discovered per year. For example, when $n = 50$ sites have been sampled from archaeological deposits spanning t = 10,000 years, $\lambda = n/t = 50/10{,}000 = 0.005$. Under these conditions, after ordering the sites chronologically, the probability distribution of time gaps is exponentially distributed (Rhode et al., 2014; Short et al., 2009; Strauss and Sadler, 1989). In other words, the probability P of observing a time gap of duration d, $P(d)$, is

$$(4.2) \qquad P(d) = \lambda e^{-\lambda d},$$

where e is the base of the natural logarithm. This equation says that the probability distribution of time gaps is dominated by short time gaps but has a long tail that incorporates rare but long gaps (fig. 4.3). In other words, most gaps will be short, but every now and then there will be a large gap.

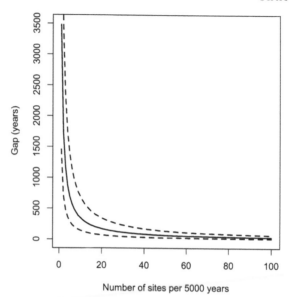

FIGURE 4.4: Median time gap as a function of the number of sites sampled, based on sampling $n =$ 100,000 time gaps from the probability distribution described in eq. 4.2 for each site sample size. Sites are sampled with probability λ from an interval of time of 5000 years (see text for details). The model assumes that gap length cannot be longer than 5000 years. The solid line represents the median gap duration, and the dashed lines represent the 25th and 75th percentile gap duration.

The median of exponential functions such as equation 4.2 is $\ln 2\lambda^{-1}$. Since $\lambda = n/t$, the median time gap in a sequence of archaeological sites is

$$(4.3) \qquad \text{median gap} = \ln 2 \left(\frac{n}{t} \right)^{-1} .$$

This equation describes how the median time gap decreases in proportion to sample size. For instance, if the number of sites sampled per 5000-year intervals (i.e., $t = 5000$) is 10, then the median time gap between the sites is 344 years (fig. 4.4). Doubling the size of the sample to $n = 20$ cuts the median gap by roughly half (174 years). Again, doubling the sample size to 40 further halves the median gap (87 years). This inverse proportional relationship between time gaps and site sample size holds for all values of site sample size and all amounts of time t.

The uniform-frequency model illustrates how a significant temporal hiatus can be observed simply as a result of sampling error (Rhode et al., 2014). Because the distribution of time gaps is exponential, long gaps, though rare, are always possible, even for a sizable sample.

The model also tells us that the sampling interval of archaeological data will improve nonlinearly with the intensity of field research in a region.

Similarly, the nature and the intensity of human activity in the past set an upper boundary on the number of sites that are available for archaeologists to sample. Even with an equal amount of field research, the sampling interval of the archaeological record of a region occupied only sporadically by small populations of hunter-gatherer tribes will always be longer than that of a region occupied continuously by large sedentary groups over thousands of years, since it will contain fewer datable sites.

Wider Sampling Universes Lead to Longer Time Intervals

The uniform-frequency model assumes that sites are sampled from an infinite time line and that there are no limits to how long gaps can be. For instance, when equation 4.3 says that the median time gap between 10 sites sampled over a 5000-year period is 344 years, it incorporates in its estimates time gaps that are longer than 5000 years (however rare they are).

In the real world, archaeologists sample sites from finite time lines—they work with sampling universes that have a finite span and duration. These sampling universes are limited by two orders of boundary.

The first-order boundary is set by the present time—time gaps cannot extend into the future. A time gap cannot be longer than the age of the oldest site from which it is calculated. For instance, the gap between a 5000-year-old site and a younger site cannot be longer than 5000 years. The second order of boundary stems from the fact that archaeologists rarely sample the global archaeological record. Rather, they collect data from a particular region and time period. Both orders of boundaries influence the sampling interval of archaeological data by setting an upper limit on the amount of time that can separate two samples.

When sampling takes place on a finite time line rather than on an infinite time line, the probability distribution of time gaps is not exponential, as is the case in the uniform-frequency model, but rather follows a Dirichlet distribution (Strauss and Sadler, 1989). The properties of the Dirichlet distribution are more complicated to describe than those of the exponential distribution, so I used simulations to study the impact of the duration of the sampling universe on time gaps.

I took samples of $n = 100,000$ pairs of sites from sampling universes that varied in duration from 2 to 40,000 years and measured the time gaps within each pair. Plotting the gaps against the age of the older site in the pair yields a wedge-shaped distribution (fig. 4.5). The distribution is wedge shaped because the age of the older site limits how long the gaps can be. For example, if the older site is 30,000 years old, the gap between the sites cannot be longer than

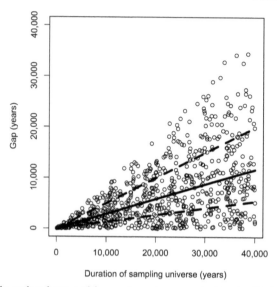

FIGURE 4.5: The median duration of the sampling universe covaries linearly with time gaps. For any given duration of sampling universe ranging from 2 to 40,000 years, $n = 100,000$ pairs of sites are sampled from a uniform-frequency distribution, and the time gap between them is calculated (see text for details). The solid line represents the median time gap, and the dashed lines represent the 25th and 75th percentile gap duration. For visibility, only $n = 750$ data points are plotted on the chart.

30,000 years. The time gap between archaeological sites varies linearly with the duration of the sampling universe (fig. 4.5). For instance, the median time gap between pairs of sites sampled from a 1000-year-long universe is 293 years, and the 25th and 75th percentile gaps are 134 and 499 years, respectively. Doubling the duration of the sampling universe to 2000 years leads to a median time gap that is twice as long, 585 years (268 and 998 years for the 25th and 75th percentiles); increasing it tenfold to 10,000 years leads to time gaps that are ten times longer, 2924 years (1337 and 4996 for the 25th and 75th percentiles).

These results apply to both types of sampling-universe boundaries. The duration of the sampling universe can represent the first-order boundary (the boundary set by the present time) if we assume that it represents the age of the oldest site that could possibly be included in a sample. For instance, a duration of 2000 years may represent the case where archaeological sites are sampled from a universe that stretches from 2000 BP to the present. But the same results can also capture the second type of boundary—that is, the situation in which sites are sampled from a period of 2000 years, for instance, 8000–6000 BP.

Thus, archaeologists' research interests determine, in part, the sampling interval of their data. Archaeologists working on short-lived phenomena, such as the spread of a particular ceramic style, will enjoy better sampling intervals than archaeologists working on longer-lived phenomena, such as the Middle Stone Age of Africa, because they are sampling from a shorter time line. This is true even for archaeologists sampling from the same regional record: the archaeologist who collects both preceramic and ceramic-period remains in a region will have to contend with longer time gaps than the archaeologist working in the same region but who analyzes only the ceramic-period sites.

Age-Biased Preservation Loss Leads to Shorter Time Gaps
(All Other Things Being Equal)

Preservation loss further alters the sampling interval of the archaeological record. First, preservation loss sets an upper limit on the size of archaeological samples by affecting the size of the population of physical finds. If only 25 archaeological sites have been preserved in a region, the sample cannot contain more than 25 sites. More counterintuitive, however, is the effect of age-biased preservation loss. Age-biased preservation loss, whereby older sites are more likely to have been destroyed than younger sites, creates a "pull of the recent" in the archaeological record that results, all other things being equal, in shorter time gaps between sites. The "pull of the recent" skews the temporal frequency distribution of archaeological remains toward younger ages. As a result, when a site is discovered, it is more likely to be of younger age than of older age. Now, for a sampling universe of any given duration, age-biased loss means that the age of the sites in a sample will tend to aggregate toward the younger end of the sampling universe and, because of that, will have shorter time gaps between them. In contrast, in the absence of age-biased preservation loss, the age of the sites will be uniformly distributed within the sampling universe and thus have longer time gaps between them. The stronger age-biased preservation loss is, the shorter the gaps in the archaeological record will be, all other things being equal.

To explore the effect of age-biased preservation loss on time gaps, I replicated the simulation presented in the previous section. But instead of sampling pairs of sites from a uniform-frequency distribution, I sampled them from a distribution affected by the age-biased taphonomic loss process identified by Surovell et al. (2009) in the terrestrial volcanic record. The probability of sampling a site of age q, $P(q)$, is

$$(4.4) \quad P(q) = \prod_{t=1}^{q} 1 - \frac{1.3925309}{2176 + t}.$$

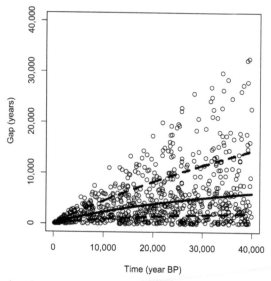

FIGURE 4.6: Age-biased preservation loss leads to shorter time gaps. For any given duration of sampling universe ranging from 2 to 40,000 years, $n = 100,000$ pairs of sites are sampled from a uniform-frequency distribution, and the time gap between them is calculated (see text for details). The solid line represents the median gap duration, and the dashed lines represent the 25th and 75th percentile gap duration. For visibility, only $n = 750$ data points are plotted on the chart.

The rightmost term of the equation is the ρ (eq. 4.1) as parametrized by Surovell et al. 2009). It represents the probability that a site is lost during year t after its creation. Thus, the term to the right of the product equation is $1 - \rho$, the probability that a site survives year t. The whole equation specifies the probability of finding a site of age q, which is the probability that it has survived every single year from the moment of its creation ($t = 1$) to the present ($t = q$).

As before, I sampled $n = 100,000$ pairs of sites from sampling universes of duration that varied from 2 to 40,000 years. Here, however, I assumed that all the sampling universes had for an upper limit the present, so that a 5000-year-long universe spans from 1 to 5000 BP, and a 10,000-year-old universe spans from 1 to 10,000 BP. This assumption allows us to compare the results of the simulation directly with the results of figure 4.5 and isolate the effect of preservation loss from that of the duration of the sampling universe.

The resulting time gaps form a wedge-shaped distribution (fig. 4.6) that looks much like that of figure 4.5. The difference between the two figures, however, is that the distribution with age-biased preservation loss is heavy at the bottom: gaps cluster at the bottom of the chart, where gaps are short. As a result, the median time gap between pairs of sites is shorter. For instance, the median time gap when sites are sampled from a sampling universe that spans from 1 to 2000 BP is 569 years, whereas it is 585 years without preservation loss.

The longer a sampling universe is, the more pronounced the effect of age-biased loss is. In the case of the taphonomic loss rate in the terrestrial volcanic record (Surovell et al., 2009), the effect becomes important when the duration of the sampling universe is on the order of 10^3 years and falls within the last 20,000 years (fig. 4.7). The effect is most pronounced when the younger limit of the sampling universe is the present (0 BP). With age-biased preservation loss, the median gap between pairs of sites drawn randomly from a universe spanning from 0 to 1000 BP (the dotted line in fig. 4.7) is about 1% shorter than what it would be in the absence of preservation loss (290 vs. 293 years). But if the sampling universe is longer and extends from 0 to 5000 BP (the dashed line in fig. 4.7), the median time gap is 11% shorter (1319 vs. 1473 years). Finally, with a sampling universe that extends from 0 to 10,000 BP (the solid line in fig. 4.7), the median time gap is 20% shorter than what it would normally be (2349 vs. 2924 years).

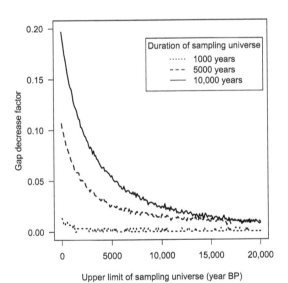

FIGURE 4.7: Decrease factor in median time gap caused by age-biased preservation loss process, assuming three sampling-universe durations: 1000, 5000, and 10,000 years. The median values are calculating from $n = 1{,}000{,}000$ samples of time gaps between pairs of sites. The age on the x-axis corresponds to the upper (i.e., younger) limit of the sampling universe. For instance, at 0 BP on the x-axis, a 1000-year sampling universe spans from 0 to 1000 BP. The median time gaps for sampling universes of 1000, 5000, and 10,000 years without age-biased preservation loss are 293, 1473, and 2924 years, respectively (fig. 4.5).

Shifting the upper limit of the sampling universes back in time decreases the effect of age-biased preservation loss. For example, the gap decrease factor changes from 11% to less than 2% when the 5000-year-long sampling universe span is moved from 0–5000 BP to 10,000–15,000 BP (fig. 4.7).

At first glance, the result that preservation loss decreases time intervals is at odds with archaeologists' intuition that the quality of the archaeological record decreases the farther we go back in time. But it is not. First, the result does not mean that preservation loss augments the quality of archaeological data. After all, although preservation loss may improve the time gaps between our observations, it comes at a cost, that of undersampling the older portions of the sampling-universe record. Second, preservation loss has consequences that are not captured in the simulation. Preservation loss creates discontinuities in the archaeological record above and beyond those predicted by the age-biased process observed in the terrestrial volcanic record. For instance, the geological deposits associated with a time period in a region may have been completely eroded away, creating a larger gap in the archaeological sequence than predicted by the model. Second, age-biased preservation loss reduces time gaps between sites *at equal sample size*. This is an important caveat, because preservation loss decreases the frequency of archaeological material (see above). When preservation loss has made archaeological remains rarer, archaeologists may have to increase the duration of their sampling universe to capture an adequate sample. Archaeologists expand their sampling universe when they study older time periods in part so that they can tally a workable sample size. For instance, an archaeologist working on the colonial period of the United States can easily collect dozens of sites from the eighteenth century. A Paleolithic archaeologist, however, is unlikely to ever find two contexts that date from the same century. This joint effect of smaller sample size and wider sampling universe will generally be strong enough to counteract any positive effects that age-biased loss may have on the sampling interval of archaeological data. Thus, while age-biased preservation loss leads, all other things being equal, to shorter time gaps, it will in practice lead to longer time gaps.

Putting It All Together: The Sampling Interval of the Archaeological Record Ranges from 10^0 to 10^3 Years

Now let's put it all together. What kind of time gaps should we expect to see in the archaeological record when samples of n sites are taken from a universe of duration t and when sites were lost to preservation at the age-biased rate ρ?

Let us assume that archaeologists, to obtain adequate samples, expand their sampling universe as they study older time periods. Realistically, we can imagine that the duration of sampling universes is equal to 50% of the oldest occupation age that an archaeologist would consider excavating. For example, if sites have to be no older than 1000 BP to be of interest to an archaeologist, the sampling universe would be 500 years long (500–1000 BP). Similarly, if the lower age limit is 20,000 BP, the sampling universe is 10,000 years long (10,000–20,000 BP). This number is not arbitrary: it corresponds to the median ratio between the temporal scope and the earliest date in the archaeological literature (chapter 5). Let's also assume that sample size varies from 2 to 100, and that the rate of preservation loss ρ identified in the volcanic record (Surovell et al., 2009) is valid beyond 40,000 BP.

Under these conditions, the expected temporal sampling interval of the archaeological record of the last 100,000 years varies from 10^0 to 10^4 years (fig. 4.8). Short time gaps on the order of 10^0 years are extremely rare and are confined to the recent past and large sample sizes. Similarly, long time gaps of 10^4 years are restricted to time periods older than 70,000 years and samples of fewer than 5 sites. The model predicts that, in most cases, the expected median time gap will be on the order of 10^1 (i.e., 10–99 years) to 10^3

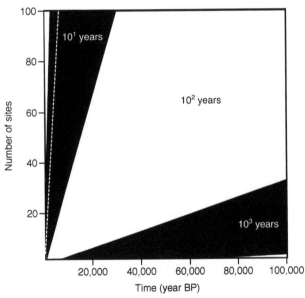

FIGURE 4.8: The expected temporal sampling interval of archaeological data ranges from 10^0 to 10^3 years. For any given point in time (x-axis) and site sample size (y-axis), $n = 1000$ samples are taken, with replacement, from a sampling universe equal to 50% of the point in time. The chart shows the parameter space under which the median time gap among the samples is on the order of magnitude 10^0 (white area on the upper-left side), 10^1, 10^2, 10^3, and 10^4 years (the small white area in the lower-right corner). The dashed white line indicates the boundary for median time gaps of one human generation (20 years).

(i.e., 1000–9999 years). What is more, the model predicts that under the vast majority of conditions, the typical time gap will be longer than one human generation (about 20 years). Intervals shorter than a human generation are not expected in the archaeological record beyond around 7000 BP or when sample sizes are smaller than 10.

The upper-right region of figure 4.8 is unlikely to ever be explored. Both preservation and observational loss are always pulling the sampling interval of any dataset to the bottom of the chart by decreasing the number of sites available. Increasing the duration of the sampling universe can only do so much to compensate for loss, and archaeologists working on the oldest portions of the record will always have to contend with small sample size, no matter what.

A Wider Sampling Universe Leads to Longer Space Intervals

What about the spatial sampling interval of the archaeological record? Even though archaeologists privilege time over space, the spatial distance between archaeological units is as much a source of underdetermination as the time gap between them. It is thus equally important to build a theory of the determinants of spatial sampling interval. For instance, how is spatial sampling interval affected by sample size and the spatial dimensions of the sampling universe?

The relationship between spatial gaps, sample size, and the dimensions of the sampling universe is simpler than is the case for time gaps. Imagine the simplest scenario in which the sampling universe is a square area and archaeological sites are distributed randomly within it. How does the expected distance between these archaeological sites vary as a function of the size of the square? The solution to this problem has already been worked out by mathematicians: the median pairwise distance between each and every pair of sites converges to 0.512 of the side length of the square (Weisstein, 2015). What that means is that the median distance between archaeological sites in a 100 × 100 kilometer square is 51.2 kilometers. Similarly, the median distance between sites that are randomly distributed in a 1000 × 1000 kilometer area is 512 kilometers. Sample size matters little here. As sample size increases, the distance rapidly converges to 0.512 of the side length of the square, after which adding more sites does not lead to shorter median pairwise distances.

There is thus a trade-off between time sampling interval and spatial sampling interval. Increasing the spatial area of the sampling universe may be seen as a panacea to long time intervals. After all, a wider sampling universe increases the chances of discovering a site that will fill the time gap between two other units. But increasing the area of the sampling universe comes at a cost—that of an increase in the spatial distance between archaeological units. Any gains in improving the temporal sampling interval of the data by increasing the spatial scope would be accompanied by a worsening of the spatial sampling interval of the data.

THE TEMPORAL RANGE OF CULTURAL TRAITS IS UNDERESTIMATED BECAUSE OF LOSS

One of the most important systematic biases in the archaeological record is that it always underestimates the true range of cultural traditions. Because of loss, it is always unlikely that the true first (and true last) instance of a cultural trait is ever found. As a result, temporal ranges in the archaeological record are always shorter than true ranges (fig. 4.9). This is a problem, as the date of appearance of different archaeological phenomena (e.g., when fire was first manipulated) and the timing of historical events (e.g., when humans first enter Australia) shape our understanding of the past in profound ways. In fact, many archaeological hypotheses are tested by translating them in terms of relative timing between variables, with the causal variable appearing before its consequent. For instance, if large-scale irrigation systems caused the rise of complex societies, then they should precede the first manifestations of complex societies.

We underestimate temporal ranges of cultural traits because of three related reasons: (1) sampling errors, which are exacerbated (2) by the fact that traits are relatively infrequent at the boundaries of their temporal range and (3) by preservation and observational loss.

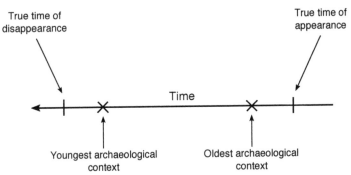

FIGURE 4.9: The archaeological record underestimates the true temporal range of cultural traits. The earliest known instance of a cultural trait associated with a cultural tradition is younger than the true time of appearance of the tradition. Similarly, the latest known instance of the trait is older than the true time of disappearance of the tradition.

Archaeologists underestimate temporal ranges primarily because of sampling errors. Archaeological datasets always represent a limited sample drawn from the larger set of archaeological sites that are available to be discovered, known and unknown. The smaller this sample is, the less likely it is to incorporate the earliest and the latest occurrence of a cultural trait.

This problem is compounded by the fact that traits are infrequent at the boundaries of their temporal range. Typically, the temporal frequency distribution of a trait is either concave or S-shaped. A concave distribution emerges when the population of a new trait waxes and wanes, producing the iconic battleship curves when seriated. S-shaped distributions are characteristic of traits that are maintained in a population for a long period of time because of their functional value. For instance, ceramics and the making of fire have rarely disappeared from the human cultural repertoire after their appearance. In the case of concave distributions, the frequency of a trait is at its lowest around its time of appearance and disappearance; in the case of S-shaped distributions, it is at its lowest at the time of appearance.

This makes the discovery of archaeological material dating to these periods of low frequency a rare event. In fact, the archaeological record most likely presents rises and drops in popularity rather than the actual times of appearance and disappearance.

In addition, the overall abundance of a trait influences the extent to which temporal ranges are underestimated (McKinney, 1991). The temporal range of traits that are rare in the dynamic context will be more greatly underestimated than the range of abundant traits, because rare traits are less likely to appear in the archaeological record and be discovered by archaeologists. For instance, houses are built less frequently than ceramic cooking vessels are produced. Because of their lower baseline frequency in dynamic contexts, the temporal range of any particular house design may be recorded as shorter than that of a ceramic style.

The spatial range of traits also interacts with how we observe their temporal range. Of two traits with the same temporal range, the one with a small geographic range will be recorded as having a shorter temporal range, as it is less likely to be sampled by archaeologists (McKinney, 1991). In addition, the frequency of a trait may be highest near the center of its geographic range and lowest at its margin. Range estimates based on archaeological deposits near the margin will thus underestimate the true range more greatly than deposits near the center (McKinney, 1991).

Years ago, I built a simulation to study the impact of sample size on the reconstruction of cultural histories (Perreault, 2011). I used the simulation to describe how sample size affects the accuracy and the precision of our estimates of six aspects of culture history: (1) date of appearance; (2) date of disappearance; (3) date and (4) magnitude of the peak in popularity; (5) rate of spread; and (6) rate of abandonment. By accuracy, I am referring to how close the mean of repeated independent measurements is to the true value of a parameter. By precision, I mean the extent to which repeated independent measurements vary around their mean. What I found is that very small samples yield both inaccurate and imprecise estimates of earliest and latest dates. Take a cultural trait that lasted 500 years, appearing at year 1 CE and gone by 500 CE. In between, it increased and decreased in popularity following a bell-shaped function that peaked at 250 CE (fig. 4.10, left). Estimates of when the trait first appeared, based on a small sample, are vastly inaccurate. When the sample size is 1 (i.e., only 1 site is known), the median estimate of the earliest age is 250 CE (because the temporal frequency distribution of the trait is symmetrical). Estimates are also imprecise: 50% of the time, the estimate of the time of appearance of the trait falls between 184 and 317 CE (fig. 4.10, right). Then, as more sites are added to the sample, accuracy grows asymptotically. With 10 sites, the expected estimate of the time of appearance is still off by more than a century—a fifth of the actual temporal range of the trait. With 20 sites, it is off by 75 years. With 100 sites, a very large sample given the duration

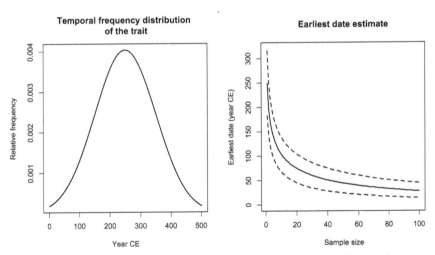

FIGURE 4.10: Estimates of the date of appearance of a cultural trait as a function of site sample size. A cultural trait appears in the year 1 CE, peaks in popularity in 250 CE, and disappears by 500 CE, yielding a temporal frequency distribution of a trait that is bell-shaped (left). For each site sample size from 1 to 100, $n = 100,000$ dates are drawn, with replacement, from the temporal frequency distribution of the trait, and the earliest date in each sample is tallied. The median earliest date (solid line) and the 25th and 75th percentile earliest date, as a function of sample size, are plotted on the right.

of the tradition (1 site per 5 years), the estimates are off by 27 years. Thus, even though two traditions have the same temporal range, one will appear to have a shorter life span than the other if it is known from a smaller sample of sites. Precision improves more slowly with sample size than accuracy does (Perreault, 2011). Among samples of 20 sites, 50% of the estimates will fall between 45 and 104 CE, a spread of 59 years (fig. 4.10, right). And with a sample of 100 sites, 50% of the estimates will fall between 14 and 44 CE, a spread of 30 years.

When the temporal frequency of a trait is asymmetrical, however, the time of appearance and the time of disappearance require different sample sizes in order to be estimated accurately and precisely (Perreault, 2011). An asymmetrical distribution could arise, for example, when the history of a trait is marked by a long period of adoption but a short period of abandonment. In this example, estimating the time of appearance accurately would require a larger sample than estimating the time of disappearance.

Preservation loss exacerbates the underestimation of temporal ranges by further depleting the already narrow tails of the temporal frequency distribution. When preservation loss is not age biased (young and old sites are equally likely to be destroyed), its effect is equivalent to a reduction of sample size (Perreault, 2011). But age-biased preservation loss lowers the accuracy of time-of-appearance estimates, while increasing the accuracy of time-of-disappearance estimates. By shifting the weight of the temporal frequency distribution away from the past and toward the present, age-biased loss decreases the likelihood of sampling data that are close to the time of appearance (Perreault, 2011). Thus, in general, archaeologists have better estimates of when a tradition disappears than when it appears.

THE SIGNOR-LIPPS EFFECT

Loss makes sudden cultural change appear gradual. The effect is known in paleontology as the *Signor-Lipps effect*, after the work of Signor and Lipps (1982), who recognized that the sudden extinction of multiple taxa would appear in the fossil record as a smeared-out, sequential series of extinctions because of incomplete preservation.

Cultural change can be abrupt, such as when a particular way of life is abandoned after an environmental crisis, a societal collapse, or a demographic crash. And as with the fossil record, errors in the estimation of the time of disappearance of various traits will result in apparent gradual change (fig. 4.11). This is because the most abundant and preservable traits will persist, archaeologically, close to the boundary event, whereas the other traits

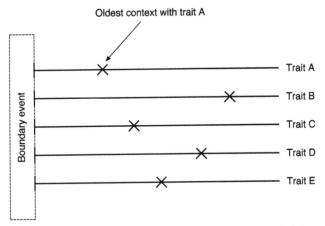

FIGURE 4.11: The Signor-Lipps effect makes sudden cultural change appear gradual. Errors in estimation of the times of disappearance of different traits result in apparent gradual disappearance, whereas the traits disappeared at the same time owing to a boundary event.

will disappear before, depending on their abundance and preservability. The Signor-Lipps effect also operates on origination events. The sudden appearance of several new traditions, triggered, for instance, by the massive arrival of migrants in a region or an imperial conquest, will look gradual and sequential because of loss.

The Signor-Lipps effect explains why, for example, the archaeological record of *Homo sapiens* predates its fossil record, because stone tools preserve better than bones (Morgan and Renne, 2008). Thus, the discrepancy between the timing of the emergence of modern human behavior in the archaeological record and the appearance of anatomically modern humans in the fossil record may not need a special explanation beyond that of the Signor-Lipps effect. Similarly, gradual change in the archaeological record should not be considered evidence against abrupt, catastrophic change—nor should it be considered evidence for gradual change. For example, the gradual extinction of megafauna at the end of the Pleistocene period (Wroe and Field, 2006) or the gradual cessation of monument building during the collapse of Maya society (Gill et al., 2007) may have more to do with the Signor-Lipps effect than with the actual abruptness of these events.

LOSS SLOWS DOWN APPARENT RATES OF CHANGE

One of the dominant features of the global archaeological record is that the pace of change in material culture appears faster in more recent periods than it does in older ones. This feature is explained, at least in part, by the effect of loss on rates of change.

Paleontologists have long been interested in how fast species evolve. One of the metrics that paleontologists use to measure rates of change is the "darwin" (d), a standardized unit of change in factors of e, the base of the natural log, per millions of years (Haldane, 1949):

$$d = \frac{\ln x_2 - \ln x_1}{\Delta_t},$$

where x_1 and x_2 are the mean trait value at time 1 and 2 (e.g., the mean body size in a population at time 1 and 2), ln is the natural logarithm, and Δ_t is the time interval between x_1 and x_2 in millions of years. (The time scale of darwins is in millions of years because when J. B. S. Haldane devised it in 1949, he assumed that natural selection operated over such long time scales. Today we know that this is not true and that natural selection can lead to significant morphological change over just a few generations.) The mean values of the trait are scaled logarithmically to control for the size magnitude of traits—this makes an increase from 1 to 2 centimeters equivalent to a growth from 100 to 200 centimeters, for instance.

Paleontologist Philip D. Gingerich made a puzzling discovery while studying a large collection of rates of change in the fossil record: he found that rates of evolutionary change are inversely correlated with the time interval over which they are measured. Figure 4.12 reproduces the chart published by Gingerich in the *Science* paper in which he reported his finding (Gingerich, 1983). You can see that the rates calculated over short time intervals (the left side of the chart) are much faster than the rates calculated over longer time intervals (the right side of the chart). This is weird. When you calculate a rate, when you divide an amount of change by time, what you are effectively trying to do is to get rid of the effect of time intervals. Why are rates of evolution dependent on time intervals?

Rates are dependent on time intervals for two reasons, both stemming from the fact that the process of biological evolution operates over shorter time scales than the typical sampling interval of the fossil record (Gingerich, 1983). First, as time interval increases, it becomes increasingly more likely that the net rate observed is in fact averaged over several disparate rates and evolutionary reversals and that, as a result, rapid change can be observed only over short time intervals (fig. 4.13). Second, because of the effect of stabilizing selection or functional constraints, morphologies often reach some

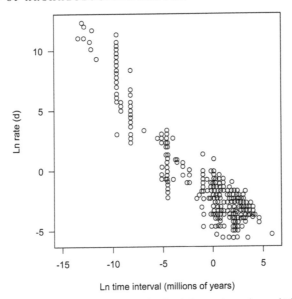

FIGURE 4.12: Rates of evolution are inversely correlated with the time interval over which they are measured. Rates of change from the fossil record, measured in darwins (d), are plotted against time intervals in millions of years, on a natural log scale. Data from Gingerich 1983 made available by author.

evolutionary stasis and consequently undergo change only rarely (fig. 4.14). Thus, when time intervals are longer than the time it typically takes for evolution to reach evolutionary stasis, rates of change will be slow compared with rates calculated over shorter time intervals.

These two factors explain why, as we go back in time, the pace of biological evolution appears to slow down, and eventually, if we go far enough back in time, what we see is mostly stasis. What is more, the two factors also explain why rates of change in nonbiological systems are also inversely correlated with the time span over which they are measured, such as rates of sedimentation (Sadler, 1981) or of change in land surface elevation (Gardner et al., 1987).

Cultural rates of change in the archaeological record show the same inverse relationship with time interval. In a previous study (Perreault, 2012) I analyzed 573 rates of cultural change that I compiled from the archaeological literature (see fig. 7.5). The rates, measured in darwins, represent change in the dimension of various technologies observed in the archaeological record of North America, such as changes in Anasazi pit structure depth and in the size of printer type block in Annapolis (Perreault, 2012). What is more, because time intervals increase as we go back in time, these rates of change are also inversely correlated with the age of the material (Perreault, 2012). This

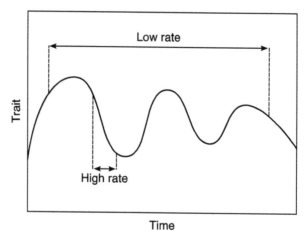

FIGURE 4.13: Reversals in the value of a trait can generate an inverse correlation between rates of change and time intervals. Rapid change in body size can be observed only over short time intervals, whereas, when observed over long time intervals, the net amount of change in body size is averaged over several reversals, leading to slower rates of change.

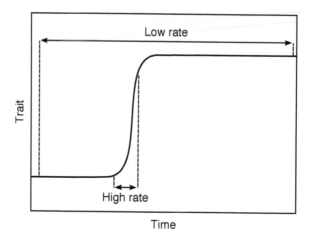

FIGURE 4.14: Stasis can generate an inverse correlation between rates of change and time intervals. Because of the effect of stabilizing selection or functional constraints, the amount of change (the numerator in the rate equation) is bounded, whereas time (Δ_t, the denominator in the rate equation) is unbounded and free to increase.

explains why the pace of change in material culture appears to slow down as we look at older time periods, from the objects that surround us in our daily lives, to the material culture at the time of our grandparents, to the material culture from 10,000 years ago. And as with the fossil record, if you look back far enough in time, what you see in the archaeological record is mostly stasis. This global pattern in the archaeological record is due to the fact that we are not observing these different periods of our history using the same observational time scales.

The decay of information that results from preservation and observational loss severely limits our capacity to control for covariates. This, in turn, means that we may not be able to shield our research from false-negative or false-positive results.

Controlling for covariates is paramount in social sciences. Humans are complex creatures, and their actions are shaped by a myriad of factors that operate at different scales and that interact with each other. In fact, humans are so complex that even the most reductionist approaches to human behavior recognize that to test any hypotheses, one needs to control for a long string of covariates, such as age, gender, number of offspring, or group size. These covariates need to be controlled for because they can all influence the analyst's target variable, and by leaving them unchecked, the analyst risks reaching a false conclusion.

Take the example of the size of tool assemblages at !Kung San camps (table 3.3). All other things being equal, the size of a tool assemblage should increase with occupation span, and an archaeologist excavating the !Kung San camps recorded by Yellen (1977) may find it reasonable to treat the size of tool assemblages as a marker of occupation span. Yet, occupation span and the number of tools left at these camps are uncorrelated (Shott, 1989b). Indeed, the variation in assemblage size is largely driven by variation in the use-life of objects, a variable that is usually inaccessible to archaeologists. Controlling for tool use-life would lead to very different estimates of occupation span than an analysis in which tool use-life is ignored.

But, statistically, archaeologists work with their hands tied behind their backs. Of all the independent variables that would need to be controlled for to ensure that the "all other things being equal" clause that comes with a model is met, archaeologists have access to just a handful, at best. If that handful of independent variables is insufficient to study a process ethnographically, it is also insufficient to study it archaeologically.

Conclusion

The forces discussed in this and the previous chapter are all sources of underdetermination. They lead to underdetermination because they decrease the overall amount of information that is present in the archaeological record, either by decreasing the resolution, by increasing the sampling interval, or by decreasing the dimensionality of the record (table 4.1). And archaeologists can do only so much to control these forces. Some of them, like analytical lumping, can be modulated to a certain extent, but most are completely out of archaeologists' control. Their net effect is to set an absolute limit on the quality of archaeological datasets.

Because these forces are numerous and diverse, the sampling interval, resolution, and dimensionality can vary independently of each other (Behrensmeyer, Kidwell, and Gastaldo, 2000). For example, it is perfectly possible to have a cultural layer at a site that is high resolution but with low dimensionality because the occupants took most of their belongings with them when they left the site. Alternatively, an assemblage may be time averaged over centuries and yet be uniquely rich dimensionally because of good preservation. Or a

TABLE 4.1. The forces that affect the quality of archaeological data

Force	Effect on quality of data			Trend
	Sampling interval	Resolution	Dimensionality	
Discard	—	Decrease	—	Decrease with age
Reoccupation	—	Decrease	Decrease	—
Sedimentation rates	Increase	Decrease	Decrease	—
Degradation	—	Decrease	—	Increase with age
Preburial disturbances	—	Decrease	—	—
Postburial disturbance	—	Decrease	—	Increase with age
Analytical lumping	—	Decrease	—	Increase with age
Imprecision of dating techniques	—	Decrease	—	Increase with age
Unrecordable information	Increase	—	Decrease	—
Mobility, reuse, scavenging	Increase	—	Decrease	Increase with age
Burial, permanent architecture	Decrease	Increase	—	Decrease with age
Deterioration	Increase	—	Decrease	Increase with age
Fieldwork techniques	Increase	Decrease	Decrease	Increase with age

Note: These forces are discussed in chapters 3 and 4. Trend refers to changes in the intensity of the force as one goes back in time, from the youngest portion of the archaeological record to the oldest.

series of cultural layers at a stratigraphic site may have fine temporal resolution while being separated by hundreds of years from each other. Nonetheless, the review of these forces conducted here conveys how unlikely it is that any given set of archaeological observations has all that it takes—the short sampling interval, the fine-grained resolution, and all the necessary dimensions—to study short-term, ethnographic-scale processes.

Because of these forces, the archaeological record is not just incomplete but also biased. Historical trends have influenced the magnitude of some of the forces that shape the archaeological record. For example, the shift from a nomadic to a sedentary lifestyle that marked the Holocene period increased the opportunities for mixing through discard. Similarly, the arrival of permanent architecture, ceramic technology, and metalworking also marked shifts in the quality of the record, as they caused changes in archaeological visibility and preservation and allowed archaeologists to analytically lump material into shorter time periods (M. E. Smith, 1992). More importantly, perhaps, is the fact that many of these forces are time dependent—their effect increases with time—which means that the quality of archaeological data worsens the farther back we go in time. The "pull of the recent" that these time-dependent forces generate inevitably skews our view of human history, creating trends that could easily be mistaken for anthropologically meaningful signals.

5

The Quality of the Archaeological Record

The two previous chapters were concerned with the forces that shape the quality of the archaeological record. What the outcome of these forces is—what the quality of the archaeological record really is—is an empirical question. And answering this question is the first step toward solving the underdetermination problem because the answer will dictate what kind of research questions can, and cannot, be addressed archaeologically.

Here I focus on the sampling interval and the resolution of archaeological data and not so much on the scope or dimensionality. The scope is a parameter that archaeologists can vary with few constraints, and dimensionality is difficult to measure in a systematic manner across studies.

I also focus on the *expected* sampling interval and resolution of archaeological data. Obviously, there is a significant amount of variation in the quality of the record, and the quality of one's data may be very different from the quality of someone else's data. But it is against the expected quality of archaeological data that archaeologists need to calibrate their general research program.

Materials

The quality of archaeological data is a function of two things: (1) the intrinsic quality of the archaeological record as it exists in nature and (2) the decisions made by archaeologists, such as the field methods used, the nature of the analytical units created, and the dating techniques employed. The first item, the quality of the archaeological record, is outside archaeologists' control and sets an upper bound on what the quality of archaeological data can be, whatever decisions archaeologists make.

With this in mind, I used two sources of data to estimate the expected sampling interval and resolution of the archaeological record: peer-reviewed journal articles and regional databases. The sources complement each other. The data published in journal articles represent the data archaeologists use to test their hypotheses and draw their interpretations. They are the yarn from which archaeologists weave the story they tell in academic, peer-reviewed publications. They represent the quality of archaeological data after one or multiple rounds of analytical lumping and sampling of the data collected in the field. More often than not, however, more data are collected in the field than appear in publications. In that regard, regional databases that seek to represent the archaeological record of a region in an exhaustive manner are a better proxy for the intrinsic quality of the archaeological record.

JOURNAL ARTICLES

I surveyed the articles published in *American Antiquity, Current Anthropology, Journal of Anthropological Archaeology, Journal of Archaeological Research*, and *World Archaeology*, between the years 2000 and 2010, inclusively (appendix B). These five journals are prime venues for anthropological archaeology. The articles that appear in them represent the wide range of theories and perspectives that define archaeological research today. They also account for a wide range of regions and time periods.

Articles with a goal other than making inferences about the human past were excluded from this survey. These include articles concerned with the history of the discipline, philosophical debates, or the testing of new methods. In total, data from 402 journal articles were collected.

For each article, I tallied the age of every analytical unit analyzed. Obviously, the nature of these analytical units varies among articles. In some papers, the units are burials, while in others they are occupation levels, sites, or time periods. The data collected from the peer-reviewed articles thus represent the quality of the archaeological record at the hierarchical scales over which archaeologists typically construct their analytical units.

Seventy-six of the articles surveyed included more than one series of data. For instance, an article may compare the chronology of the rise of villages in four regions of the world and interpret the data pertaining to each region independently. In such cases, I separated the chronological data into four different data series, as the region is the primary level at which the author is interpreting his data. This is a conservative procedure that minimizes estimates of time interval.

In total, I collected 532 series of analytical units from 402 articles. Radiocarbon dates were calibrated using OxCal 4.1 (Bronk09) and the IntCal13 calibration curve (Reimer et al., 2013) and *terminus post quem* and *terminus ante quem* dates were excluded. All the dates were converted to calendar years before the present (BP).

I used three regional databases and extracted from them five different data-sets. Together, the five datasets represent a varied sample of regions, with a spatial scope that ranges from microregions to continents, as well as different types of populations: archaic *Homo* species, hunter-gatherers, small-scale agriculturalists, and complex societies.

Datasets 1 and 2: European Middle Paleolithic and Upper Paleolithic

The regional database PACEA Geo-referenced Radiocarbon Database (d'Errico et al., 2011) is an exhaustive collection of European Paleolithic radiocarbon dates, which range from the late Middle Paleolithic to the initial Holocene in Europe. The unit of analysis in this database is the cultural level—many entries represent the different levels from the same site. I extracted from this database two datasets, one containing the units assigned to the Middle Paleolithic period and one with the units designated as Upper Paleolithic (the levels attributed to both Middle and Upper Paleolithic, such as the Szeletian, Bohunician, and Châtelperronian, were included in both datasets). These two datasets represent the archaeological record of mobile foraging groups of Neanderthals (Middle Paleolithic) and modern humans (Upper Paleolithic). The Middle Paleolithic dataset includes 659 georeferenced levels and 551 dated levels (some levels are georeferenced but are not dated, and vice versa) coming from 184 archaeological sites. The Upper Paleolithic dataset contains 3691 georeferenced levels and 3676 dated ones collected from 702 archaeological sites.

Datasets 3 and 4: Near Eastern Natufian and Pre-Pottery Neolithic B

The Radiocarbon CONTEXT Database (Utz and Schyle, 2006) contains radiocarbon dates from the Near East ranging from the Upper Paleolithic to the Chalcolithic period. Again, the unit of analysis is the cultural level. I extracted from this regional database two datasets: one comprising Natufian units (including both Natufian and Late Natufian) and one comprising Pre-pottery Neolithic B (PPNB) units. The Natufian and PPNB archaeological record, produced by some of the earliest populations of farmers, represents the beginning of

sedentary life and food domestication. The Natufian dataset contains 93 georeferenced and dated units recovered from 17 archaeological sites, and the PPNB
dataset contains 150 georeferenced units and 147 dated ones from 21 different
sites.

Dataset 5: Valley of Mexico

The Valley of Mexico Archaeological Survey database is the outcome of an
intensive archaeological survey program that took place in the 1960s and 1970s
in the Valley of Mexico (Parsons, Kintigh, and Gregg, 1983). The units represent surface sites. Unlike with the other databases, these sites are not dated
radiometrically but assigned to different time periods (Parsons, 1974): the
Early Formative (3050–2750 BP), Middle Formative (2750–2450 BP), Late
Formative (2450–2150 BP), Terminal Formative (2150–1850 BP), Early Classic (1850–1550 BP), Classic (1850–1250 BP), Late Classic (1550–1250 BP), Early
Toltec (1250–1000 BP), Late Toltec (1000–750 BP), and Aztec (750–430 BP). I
used the entire regional database as a dataset in the analysis below. The Valley of Mexico dataset represents the rise of complex societies in a small and
intensely surveyed area. The dataset contains 2047 georeferenced and dated
units.

Sampling Interval of the Archaeological Record

MEASURING SAMPLING INTERVAL

To measure sampling interval, I began by ordering chronologically the analytical units listed within each one of the data series from journal articles or
within each regional dataset. Then, I calculated the time interval between
each and every pair of consecutive units within the same series. For example,
if the three units described in a paper are dated to 1200, 1050, and 800 BP,
I calculated the time intervals between 1200 and 1050 BP (150 years) and between 1050 and 800 BP (250 years). Many units, however, were dated to a
range of ages, such as 5000–4000 BP. When calculating the time interval between such units, I took the average of the shortest and longest possible interval between the units, in order to preserve the information about the uncertainty of these dates. For example, the shortest interval of time between
a unit dated to 5000–4000 BP and one dated to 3500–2500 BP is 500 years
(the interval between 4000 and 3500). Conversely, the longest time interval
between them is 2500 years (the gap between 5000 and 2500). This means that

the time interval between the two units could be anywhere between 2500 and 500 years, or, on average, 1500 years.

I also calculated the age of each of these time intervals, so that I could investigate the relationship between sampling interval and the age of archaeological deposits. I measured the age of an interval as the average age of the two units. Thus, the age of an interval between two units dated to 1050 and 800 BP is 925 BP. When one or both of the units in a pair are dated with a range of dates, the age was taken as the average between the oldest and the youngest possible date. For instance, the age of the interval between two units dated to 5000–4000 BP and 3500–2500 BP is the average of 5000 and 2500 years, that is, 3750 BP.

Of the 5608 time intervals calculated from journal articles, 301 (5.4%) were equal to zero. Intervals of zero happened when two units were dated imprecisely and to the same age. For instance, the interval between two units dated to "approximately 6000 years" is zero. However, it is extremely unlikely that the two units represent perfectly contemporaneous events. For that reason, the duplicates of a date within a data series were eliminated.

The spatial sampling interval could be measured in only the regional datasets. Journal articles generally do not provide the spatial coordinates of the units they discuss. In contrast, the regional databases sampled provide the location of the archaeological sites they include. However, with only the coordinates of the sites to georeference the data, it is impossible to calculate the spatial distance between units coming from the same site. Therefore, the spatial sampling interval measured here is the between-site spatial-interval scale. This is an important limitation, since archaeological research is often conducted at a within-site spatial scale.

With each regional database, I calculated the spatial distance between each unique pair of sites. The distances are as-the-crow-flies measurements, using the spherical law of cosines, in order to take into account the curvature of the earth. In the Valley of Mexico, the spatial location of the sites is rounded to the nearest meter. But in the other datasets, the locations of the sites are reported as longitude and latitude rounded to the second decimal place. Depending on where you are on the planet, the second decimal place can represent as much as 1.1 kilometer—more than enough to separate distinct archaeological sites. Because of this lack of precision, there are distinct sites that share the same coordinates. Of the 3,703,600 pairwise spatial distances tallied from the regional datasets, 3532 (less than 0.1%) were equal to zero and were eliminated from the analysis.

In total, the material I assembled contains 6509 time intervals from the peer-reviewed literature, 2046 from the Valley of Mexico dataset, 147 from the PPNB dataset, 91 from the Natufian dataset, 3675 from the Upper Paleolithic dataset, and 550 from the Middle Paleolithic dataset. It also contains 2,094,081 spatial intervals from the Valley of Mexico dataset, 231 from the PPNB dataset, 171 from the Natufian dataset, 232,221 from the Upper Paleolithic dataset, and 16,471 from the Middle Paleolithic dataset.

TEMPORAL SAMPLING INTERVAL

In Journal Articles

As I surveyed the journal articles, two types of data immediately stood out: bioarchaeological data and archaeological data from the American Southwest. Bioarchaeological studies in which the analytical units are burials had larger samples than the other archaeological studies in which the units are cultural ones. As a result, the average bioarchaeological article contained almost twice as many time intervals as nonbioarchaeological studies (24 vs. 14). This is not surprising since burials are special features: a prehistoric cemetery can contain dozens of burials, each of which is a well-defined archaeological context that can be excavated and dated individually. Burial practices influence sampling interval by leading to larger samples, which, in turn, lead to shorter time gaps between the units (chapter 4).

The archaeological record of the American Southwest also has shorter time intervals than most other regions, thanks to a combination of forces that includes a precise dating technique (dendrochronology), good preservation, high site visibility, the presence of permanent architecture that allows individual contexts within a site to be dated independently, decades of intensive research effort, and a particular focus on the last millennium of the region's prehistory.

For these reasons, the intervals from "general" contexts (i.e., not from burial contexts and from outside the American Southwest), from burial contexts, and from the American Southwest are analyzed separately. This allows for a more accurate description of the quality of the data in each one of these contexts. This is useful because bioarchaeological and archaeological data are often analyzed separately, and because most of us never deal with an archaeological record like that of the American Southwest.

Expected Time Intervals Are Greater Than One Generation   The time intervals in the journal articles vary greatly, encompassing orders of magnitude that range from 10^{-1} to 10^6 years. The time intervals in general contexts ($n = 4490$) vary from a minimum of 0.5 years to a maximum of 4,000,000 years. Their distribution is heavily skewed, with a long tail: its central tendency is

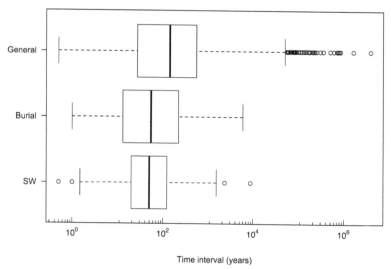

Time interval (years)

FIGURE 5.1: Boxplots of time intervals in journal articles in general contexts ($n = 4490$), in burial contexts ($n = 632$), and in the American Southwest (SW) ($n = 486$). Boxes show the median and the 25th and 75th percentiles; error bars show $1.5 \times$ IQR (interquartile range).

better captured by its median (140 years) than by its mean (5482 years). Its 25th and 75th percentiles are 27 and 550 years, respectively (fig. 5.1). The 25th to 75th percentile range is a measure of dispersion around the median and defines the range within which 50% of the values fall. Hereafter, the 25th and 75th percentiles are specified in parentheses following the median—for example, 140 years (27–550).

Time intervals can also be measured in terms of human generations instead of years. Biologically, generation time is the mean age of females at first reproduction (Charlesworth, 1994), which in the case of humans is about 20 years (Gurven and Kaplan, 2007). Generation time is thus different from life expectancy, which, at the beginning of the twenty-first century, was 66 years for humans (Gurven and Kaplan, 2007). Human generation time is a useful unit of time, because many of our theories describe processes that operate over time scales that are well within the lifetime of individuals. Thus, in terms of generation time, the median time interval in general contexts is 7 generations (1.4–27.5) long. Seventy-eight percent of the intervals are longer than 20 years (1 human generation), 55% are longer than 100 years (5 generations), and 15% are longer than 1000 years (50 generations).

In burial contexts ($n = 632$), time intervals are shorter and range from 1 to 5951 years, with a median of 54 years (13–219), or 2.7 generations (0.7–11).

Sixty-seven percent of them are longer than 20 years, 39% are longer than 100 years, and only 4% are longer than 1000 years. In a two-sample Kolmogorov-Smirnov test, the intervals in burial contexts are significantly different from those in general contexts ($D = .19$, p-value $< .0005$).

Time intervals in the American Southwest ($n = 486$) range from 0.5 to 8821 years, with a median of 50 years (20–121), or 2.5 (1–6) generations. Of these, 75% are longer than 20 years, 29% are longer than 100 years, and only 2% are longer than 1000 years. They are different from intervals in general contexts (two-sample Kolmogorov-Smirnov test, two-sided, $D = .30$, p-value $< .0005$) and from those in burial contexts ($D = .14$, p-value $< -.0005$).

Time Intervals Increase the Farther We Go Back in Time   The age of archaeological material covaries with time sampling interval. For instance, the time intervals that are on the left side of the boxes in figure 5.1 (i.e., the intervals that are shorter than the median interval) tend to be of a younger age than the ones that are on the right side of the boxes.

The time intervals vary widely in age. In general contexts, the youngest interval is 105 BP and the oldest is 4,750,000 BP, with a median age of 3219 BP (1075–8500). Units in burial contexts are younger, as burials become more elaborate and visible after the transition to agriculture. The ages of the time intervals between burials range from 156 to 13,795 BP with a median of 3400 BP (1427–5300). Finally, the intervals from the American Southwest are even younger, spanning from 300 to 12,432 BP, with a median of 880 BP (731–1200).

To examine how time intervals are affected by the age of archaeological deposits, I fitted mixed linear regression models to the intervals in general, burial, and Southwest contexts. In the three models, time interval is the outcome variable, age is a dependent variable, and the journal article from which the intervals come is a random effect (random intercept). This random effect controls for the fact that the data points that come from the same article are not independent of each other (the numbers of journal articles represented in general, burial, and Southwest contexts are 331, 26, and 49, respectively; these numbers add up to more than the total number of journal articles represented in the dataset, 402, because some articles contain data from more than one category). The data were analyzed using a Bayesian generalized linear mixed model (GLMM) approach with Markov chain Monte Carlo (MCMC) methods to estimate the parameter values using the *rjags* package in R (Plummer, 2013). The linear model is

$$\log \text{Interval} = \text{Normal}(\mu_i, \sigma);$$
$$\mu_i \sim \alpha + S_i + \beta \log Age_i.$$

With weakly informative priors:

$\sigma \sim$ Gamma$(0.001, 0.001)$;

$\alpha \sim$ Normal$(0, 0.001)$;

$\beta \sim$ Normal$(0, 0.001)$;

$S \sim$ Normal$(0, \tau)$;

$\tau \sim$ Gamma$(0.001, 0.01)$.

Figure 5.2 shows the time intervals from general archaeological contexts plotted against their age on a logarithmic scale (base 10). The solid line is the fitted regression model and has an intercept of −0.66 and a slope of 0.89. The 95% central credible interval of the slope parameter is 0.83–0.95. The fact that the central credible interval excludes zero is evidence that age has an effect on the duration of intervals, much like a p-value of less than .05.

The fact that the data can be modeled as a linear equation on a logarithmic scale tells us that time intervals increase proportionally with age rather than absolutely. For example, with an effect size of 0.89, an increase in age of 10% results in an increase in time intervals of nearly 9% ($1.1^{0.89}$). Similarly, an increase in age of 50% results in an increase in time intervals of 43% ($1.5^{0.89}$).

The linear model represents the *expected* time intervals. If we were to take a large sample of intervals of a given age, what would be the average interval in the

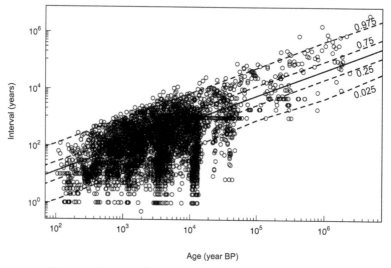

FIGURE 5.2: Time intervals in general contexts increase with time. Time intervals in general archaeological contexts are plotted against their age on a log-log scale (n = 4490). The solid line is a mixed linear regression model fitted to the data (intercept = −0.66; slope = 0.89), and the dashed lines are prediction intervals for different quantile percentile values.

sample? We can answer this question using the fitted linear model. In general contexts, the expected time interval at 100 BP is 13 years, or 65% of a human generation. By 500 BP, which roughly marks the beginning of historical archaeology in North America and other parts of the world (Pykles, 2008), the expected interval is about 54 years, or 2.7 generations. By 5000 BP, which falls within the late Neolithic period in Europe, the expected time interval is 415 years—more than 20 generations. Around the time when many groups are about to transition to agriculture, 10,000 BP, the expected time gap is 768 years long, or more than 38 generations. Sixty thousand years ago, around the time when our species may have left Africa to colonize the rest of the world (Mellars, 2006), the expected time interval is 3769 years, or 188 generations. Finally, at 150,000 BP, when our species may have been undergoing significant behavioral changes, the sampling interval of the record is 8500 years, more than 425 generations.

In burial contexts, the expected time intervals also increase with age but at a much slower pace than in general contexts (slope = 0.18; fig. 5.3), suggesting that burial contexts are robust against many of the forces that shape the record and that strength increases with time. The model shows that at 100 BP, 500 BP, 5000 BP, and 10,000 BP, the expected time intervals are 47 years, 63 years, 97 years, and 109 years, or 2.4, 3.15, 4.9, and 5.5 generations, respectively.

The American Southwest has the intervals that increase the most rapidly with age: the slope of the linear model fitted to the data is 1.02 (fig. 5.4). This means that on a log scale, time intervals increase roughly linearly with age: a 10% increase in age results in an increase in time intervals of about 10%. At 100 BP, 500 BP, 5000 BP, and 10,000 BP, the expected time intervals are 44, 62, 99, and 114, respectively. This is equivalent to intervals of 2.2, 3.1, 5, and 5.7 generations.

These values are similar to the predictions of the model discussed in chapter 4 that link site sample size and time to expected time interval (fig. 4.8). Using only the number of analytical units in the study and their age as input, the model accurately predicts the order of magnitude of the median time interval within a journal article in 263 of the 402 cases (65%) (fig. 5.5). The fact that the empirical record matches the model well supports the theory developed in chapter 4 that says that sample size and the scope and the age of the archaeological material are prime determinants of the time sampling interval of the archaeological record. The plots in figure 5.5 also show that most archaeological studies fall well within the lower-leftmost portion of figure 4.8, which corresponds to the last 10,000 years and sample sizes of 20 or less.

Thus, unless they work on very recent time periods or deal with burial contexts, archaeologists are, on average, dealing with century- and millennium-scaled time intervals (table 5.1). In general contexts, intervals of decades (10^1 years, or from 0.5 to 5 generations) are confined to the last millennium—that is,

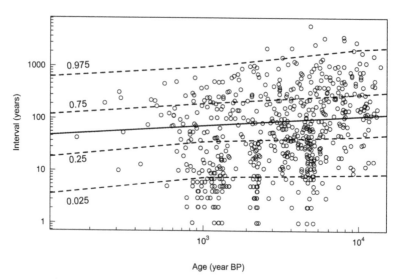

FIGURE 5.3: Time intervals in burial contexts increase with time. Time intervals in burial archaeological contexts are plotted against their age on a log-log scale ($n = 632$). The solid line is a mixed linear regression model fitted to the data (intercept = 1.31; slope = 0.18). The 95% central credible interval of the slope is 0.03–0.36. The dashed lines are prediction intervals for different quantile percentile values.

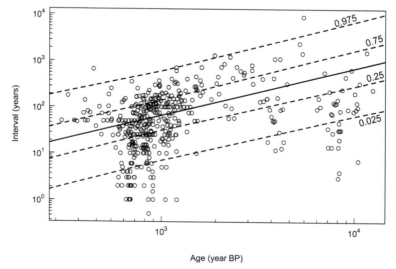

FIGURE 5.4: Time intervals in the American Southwest increase with time. Time intervals in SW contexts are plotted against their age on a log-log scale ($n = 486$). The solid line is a mixed linear regression model fitted to the data (intercept = −1.23; slope = 1.02). The 95% central credible interval of the slope is 0.76–1.29. The dashed lines are prediction intervals for different quantile percentile values.

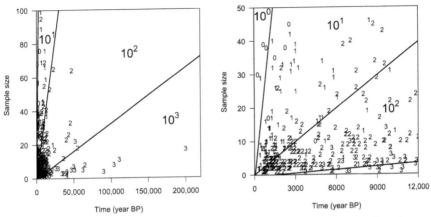

FIGURE 5.5: The order of magnitude of the median time interval in journal articles (small numbers, $n = 402$) plotted against the sample size of the study (y-axis) and the age of its oldest unit (x-axis). The data include time intervals in general, burial, and Southwest contexts. The diagonal lines and the associated large numbers define the regions over which the theoretical model presented in chapter 4 (the contour lines of fig. 4.8) predicts time intervals of a certain order of magnitude. The right plot zooms in on the lower-left portion of the left plot.

TABLE 5.1. Range of age over which different magnitudes of time intervals are expected

Interval	Age range (BP)		
	General	Burial	Southwest
10^0 years	Present to 75	Present to 0.02	Present to 0.08
10^1 years	75 to 1005	0.02–6004	0.08–5231
10^2 years	1005 to 13,456	>6004	>5231
10^3 years	13,456 to 180,181	—	—
10^4 years	180,181 to 2.4 MYA	—	—
10^5 years	>2.4 MYA	—	—

Note: Ranges are given for general, burial, and American Southwest contexts. For instance, between 1005 and 13,456 BP, the expected time intervals in general contexts are on the order of centuries (i.e., between 100 and 999 years long). The values are computed using the linear regression models presented in the text.

from 75 to 1005 BP. Between 1005 and 13,456 BP, intervals are expected to be on the order of centuries (10^2 years, or 5–50 generations). From 13,456 to 180,181 BP, expected intervals are on the order of millennia (10^3 years, 50–500 generations). For periods between 180,181 and 2,412,000 BP, the expected intervals are counted in tens of thousands of years (10^4 years, 500–5000 generations), and for those beyond 2.4 MYA, they are counted in hundreds of thousands of years (10^5 years, >5000 generations). In burial contexts, the expected time intervals range from decades, between 0.02 (about one week) and 6004 BP, to centuries, between 6004 BP and older. In Southwest contexts, expected time intervals range from decades, between 0.08 and 5231 BP, to centuries for material older than 5231 BP.

So far, the results discussed have been about the *expected* time intervals. But what about the variance in the data? In figures 5.2–5.4, there are, for any given age, some intervals that fall above the regression line and some that fall under it. This variance is due to all the forces described in chapters 3 and 4 that affect the quality of the archaeological record above and beyond the effect of age. In the statistical model fitted to the data, the effect of these forces is captured by σ as well as by S—that is, the distortion of the intercept of the model that is generated by having the data clustered by journal articles. What if instead of taking a very large sample of intervals and looking at their average, we were to sample just one time interval? What range of values should we expect a single new interval, taken from a previously unknown article, to fall in, given the variance in the data—that is, given the effect of age *plus* the variance σ and the effect of S? The dashed lines in figures 5.2–5.4 represent the answer to this question. The two innermost dashed lines represent the region within which there is a 50% chance that the one new data point falls. And the two outermost lines specify the region within which there is a 95% chance that the new data point falls. For instance, at 1000 BP, there is a 50% chance that a new interval has a duration somewhere between 94 and 104 years, and a 95% chance that it falls between 85 and 116 years. By 10,000 BP, the 50% and 95% prediction intervals are 352–1644 years and 82–7079 years, respectively. And by 100,000 BP, the 50% and 95% prediction intervals are 2807–6471 years and 4662–7673 years, respectively.

In burial contexts, including the statistical noise (i.e., the factors that affect the sampling interval of archaeological data above and beyond the age of deposits), there is at 1000 BP a 50% chance that an interval falls between 60 and 86 years and a 95% chance that it falls between 43 and 120 years. At 10,000 BP, the 50% and 95% ranges are 92–135 and 62–186 years, respectively.

In the American Southwest, there is a 50% chance that an interval dating to 1000 BP falls between 60 and 73 years and a 95% chance that it falls between 49 and 88 years. At 10,000 BP, the ranges are 565–870 and 330–1315 years, respectively.

In Regional Datasets

As in journal articles, time intervals in regional datasets are predominantly century and millennium scaled, and they increase with age. Figure 5.6 shows the distribution of the time intervals in each of the datasets, as well as the distribution of the age of these intervals. The Valley of Mexico dataset has the shortest time intervals, with a median of 125 years (100–150), or 6 generations (3–8). The PPNB dataset follows, with a median of 165 years (125–207), or 8 generations (6–10). Next are the Natufian dataset, with a median of 300 years (162–390), or 15 generations (8–20); the Upper Paleolithic period, with a median of 405 years (265–645), or 20 generations (13–32); and the Middle Paleolithic, with a median of 1045 years (654–1675), or 52.25 generations (33–84).

The variation in time intervals between the datasets is explained in part by how old the intervals in the datasets are. Each dataset covers a different period of human history, and the datasets overlap little in time. The sequence of datasets discussed above, ordered by their median time intervals, matches the sequence of datasets ordered by their median interval ages (fig. 5.6). The median ages of the intervals in the Valley of Mexico, PPNB, Natufian, Upper Paleolithic, and Middle Paleolithic datasets are, respectively, 875 BP (490–1550), 9440 BP (8960–9845), 12,610 BP (11,684–13,065), 22,140 BP (15,890–31,280), and 40,020 BP (34,700–43,510).

The time intervals in the regional datasets are shorter than those in journal articles, with one exception, the Valley of Mexico dataset. The average time interval in the Valley of Mexico dataset is 109 years, which is very close to the 104 years that the linear regression model fitted on time intervals in journal articles on general contexts predicts for an interval dated to 875 BP. In contrast, the mean interval in the PPNB dataset is shorter than the predicted mean by 65% (the mean interval is 208 years, whereas the predicted interval is 589 years); the Natufian interval is 40% shorter (445 vs. 742 years); the Upper Paleolithic interval is 51% shorter (546 vs. 1113 years); and the Middle Paleolithic interval is 14% shorter (1322 vs. 1730 years).

The time sampling interval of regional databases tends to be better than that in journal articles because databases have much larger samples (see chapter 4). On the one hand, the temporal scope of the regional dataset is much longer than the typical scope in journal articles. The median temporal scope in journal articles, calculated as the difference between the youngest and the oldest context, is 2174 years (637–6201). While this is longer than the temporal scope of the PPNB dataset (about 1600 years) and similar to that of the Valley of Mexico dataset (about 2500 years), it is shorter than that of the Natufian dataset (about 3200 years) and of the Upper and Middle Paleolithic datasets (both roughly 50,000 years). Since a wider sampling universe leads to longer time intervals (see chapter 4), we expect the time intervals in regional datasets to be longer than those of journal articles, not shorter. What is going on? What is happening is that regional datasets also have wide sampling universes. For instance, the Valley of Mexico has an area of about 2500 km^2

Time intervals

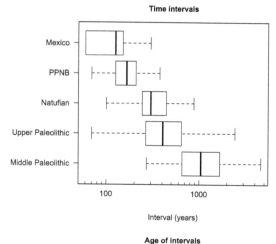

Age of intervals

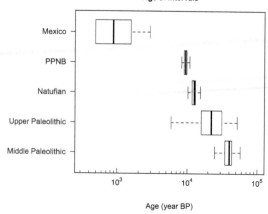

FIGURE 5.6: Time intervals in regional datasets. Top: The distribution of time intervals from the regional datasets. Boxplots show the median and 25th and 75th percentiles. Bottom: Ages of time intervals from the regional datasets. Valley of Mexico, n = 2046; PPNB, n = 147; Natufian, n = 91; Upper Paleolithic, n = 3675; and Middle Paleolithic, n = 550. Error bars show 1.5 × IQR.

(Parsons, 1974); PPNB sites are distributed over an area of roughly 320,000 km²; Natufian sites, over about 130,000 km²; Upper Paleolithic sites, over 8,850,000 km²; and Middle Paleolithic sites, over 7,920,000 km² (the spatial scope of the last four databases is estimated using the areas of the regions and the countries in which sites are found). In contrast, the spatial scope of the research presented in journal articles tends to be much smaller, such as a single site or a small region and, as a result, encompasses fewer sites. The overall effect of this is that regional datasets have larger sample sizes. For instance, the median sample size in journal articles is 8 (4–16), which, combined with a median scope of 2174, means 0.36 archaeological contexts per 100 years. In contrast, in regional databases, the number of contexts per century is always larger: 81, 6.62, 2.38, 0.41, and 0.95 for the Valley of Mexico, PPNB, Natufian, Upper Paleolithic, and Middle Paleolithic, respectively.

SPATIAL SAMPLING INTERVAL

The spatial sampling interval in the regional databases encompasses three orders of magnitude: 10^1–10^3 km (fig. 5.7). With the exception of the Valley of Mexico, where the median distance between each pair of sites is 38 km (17–66), the typical distance between archaeological sites is measured in hundreds of kilometers: 185 km (93–312) in the Natufian dataset, 502 km (256–815) in the PPNB's, 931 km (499–1511) in the Upper Paleolithic's, and 1121 km (610–1794) in the Middle Paleolithic's.

As with time intervals, age is an important determinant of the variation in spatial sampling interval between regional datasets. Sites from younger time periods tend to be closer to each other than the sites from older time periods. And once again, this correlation is because datasets from younger time periods have smaller spatial scope and because a wider sampling universe leads to longer space intervals (chapter 4). In fact, the simple model presented on page 100 predicts fairly well the median distance in each regional dataset. According to the model, the median pairwise distance between each pair of sites distributed randomly within a square area is equal to 0.512 the side length of the square.

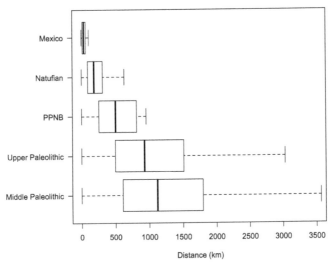

FIGURE 5.7: Spatial intervals from databases: Mexico (n = 2,094,081), Natufian (n = 171), PPNB (n = 231), Upper Paleolithic (n = 232,221), and Middle Paleolithic (n = 16,471). Boxes show the median and 25th and 75th percentiles. Error bars show 1.5 × IQR.

TABLE 5.2. Predicted and observed spatial sampling interval in regional databases

Database	Predicted spatial interval (km)	Observed spatial interval (km)
Valley of Mexico	26	38
Natufian	185	185
PPNB	289	502
Upper Paleolithic	1523	931
Middle Paleolithic	1441	1121

Assuming that the area covered by each regional dataset is a square, the model predicts median distances between the sites that are remarkably similar to the actual values (table 5.2).

Spatial scope increases with time because it is easier for archaeologists working on the younger parts of the archaeological record to amass sizable samples even when sampling from a small spatial universe. The regional datasets examined here suggest that archaeologists working on farming societies typically contend with between-site spatial intervals ranging from 10^1 to 10^2 km, whereas archaeologists working on hunter-gatherers deal with distances of 10^2–10^3 km.

Temporal Resolution of the Archaeological Record

Archaeologists have long been aware of the palimpsest nature of the archaeological record and of the fact that the temporal-resolution scale of the record varies tremendously (Meltzer, 2004), from a few hours, as is probably the case for small lithic scatters, to days (Sullivan, 1992; Wandsnider, 2008), years (Sullivan, 2008b), decades (Sullivan, 2008b; Varien and Ortman, 2005), centuries (Hosfield, 2005; Lyman, 2003; Wandsnider, 2008), millennia (Bailey and Jamie, 1997; Hosfield, 2005; Lyman, 2003; Stern, 1993, 2008; Wandsnider, 2008), and tens of thousands of years (Hosfield, 2001, 2005; Stern, 1993, 2008). What is not known, however, is the probability distribution of these different orders of magnitude. How coarse, on average, are archaeological contexts?

MEASURING RESOLUTION

Temporal resolution can be measured in two ways: as the amount of activity time represented in a context or as the total span of time represented in it. The two measures can be very different. Take a group of hunter-gatherers who visit a quarry site where they spend, once a year, about a week. Imagine that the group keeps visiting the quarry for 100 years. The amount of activity time represented at the quarry site is about 100 weeks, whereas the total span of time represented in it is 100 years. Both measures of temporal resolution are useful and complementary. An assemblage that mixes 100 weeks' worth of human activity time accumulated over a period of a century does not contain the same kind of information as an assemblage that mixes the same amount of human activity time accumulated over a millennium.

Unfortunately, the archaeological record often underdetermines activity time. Activity time is difficult to measure in the field and requires a thorough understanding of a site's deposition history, multiple chronometric dates, and the presence of taphochronometric indicators such as artifact accumulation (Varien and Ortman, 2005), the shape of hearths (see Wandsnider, 2008, table 5.1), or assemblage composition (Surovell, 2009). Often, we are able to determine only the total time span, which, unsurprisingly, is the most common method used by paleontologists to estimate the duration of temporal mixing in the fossil record (e.g., Flessa and Kowalewski, 1994; Kowalewski and Bambach, 2003). In practice, total time span is calculated as the difference between the age of the youngest and the age of the oldest dated samples in a unit.

Of course, measuring the total span of time can be done only when at least two dates are associated with the same archaeological context. Units that are associated with a single date cannot be assumed to be fine-grained Pompeii-like snapshots. Dating, after all, is expensive, and archaeologists sometimes date only one sample per context. In addition, a context may be heavily time averaged and yet contain only one specimen of datable material (e.g., only one fragment of charcoal). For these reasons, archaeological units that are associated with a single date—which includes most of the units that appear in journal articles as well as all those tallied in regional datasets—were eliminated from my analysis.

Burials were also eliminated but for a different reason. Burials represent, by nature, a very brief moment in time, and even though not a single burial discussed in the journal articles is associated with more than one date, it is safe to assume that they all have very fine temporal resolutions. The results discussed below thus represent the temporal resolution of units in general contexts and from the American Southwest.

For the units associated with more than one date, I calculated the difference between the mean of the youngest age and the mean of the oldest age. In the cases where the age of a unit is a culture time period, I calculated the time span as the difference between the lower and the upper bound of the time period. For instance, the time span of a unit dated to a time period spanning from 5000 to 4000 BP is 1000 years.

As with sampling interval, I also tallied the age of the units as the midpoint between the youngest and the oldest date. In total, I collected 1015 measurements of resolution from the journal articles, coming from 165 sources. Of these, 818 come from general contexts and 197 from the American Southwest.

TEMPORAL RESOLUTION

The temporal resolution of the archaeological units analyzed varies in magnitude from 10^1 to 10^3 years. The distribution is skewed, with a long tail (fig. 5.8). The resolution of the 818 units from general contexts ranges from 14 to 394,000 years, with a median of 400 years (187–1000), or 20 generations (9–50). More than 99% of them are longer than 1 generation ($n = 815$), 82% ($n = 674$) are longer than 5 generations (100 years), and 20% ($n = 160$) are longer than 50 generations (1000 years).

Archaeologists working in the American Southwest enjoy finer resolutions. The resolution of the 197 units from the Southwest ranges from 10 to 2050 years, with a median of 100 years (40–199), or 5 (2–10) generations. Ninety-six percent of the units have a resolution longer than 1 human generation ($n = 189$), 43% are longer than 5 human generations ($n = 84$), and only 1.5% are longer than 50 generations ($n = 3$).

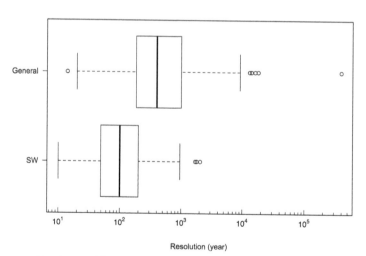

FIGURE 5.8: Distribution of the temporal resolution of archaeological units in journal articles in general contexts ($n = 818$) and the American Southwest ($n = 197$). Boxes show the median, 25th and 75th percentiles. Error bars show 1.5 × IQR. In a two-sample Kolmogorov-Smirnov test, two-sided, $D = .498$ and p-value < .0005.

Temporal Resolution Becomes Coarser with Time

As with sampling interval, the age of deposits has an important effect on the resolution of archaeological data. The units in the sample range in age from 45 to 233,000 BP, with a median of 3162 BP (1300–5467). The units in Southwest contexts range from an age of 225 to 4050 BP, with a median of 913 BP (763–1140).

This variation in age accounts for some of the variation in resolution. Fitting the same statistical model as the one used for sampling interval above, but with resolution as the outcome variable, I found that age does have an effect on resolution.

In general archaeological contexts, the expected resolution for a given age, on a log-log scale, is given by the linear model with an intercept of 0.58 and a slope of 0.61. The 95% central credible interval of the slope parameter is 0.52–0.85 (fig. 5.9).

Resolution decreases with age at a rate of 0.61, which means that for any given proportional increase in age, the total amount of time represented in a unit increases by more than half that amount. For instance, a 10% increase in age leads to an increase in resolution of 6%, and a 50% increase in age translates to a 28% increase in resolution.

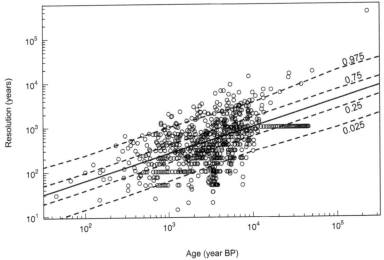

FIGURE 5.9: The resolution of the archaeological record decreases with time. The resolution of the units in general contexts ($n = 818$) is plotted against their age on a log-log scale. The data points are not all independent measurements but are grouped by journal articles ($n = 135$). This is the reason why some data points are distributed evenly horizontally in the plot: they represent sequences of time periods of the same duration. This covariance structure in the data is taken into account by fitting a hierarchical linear mixed model with the article from which the data come as a random effect (random intercept).

TABLE 5.3. Range of ages over which different orders of magnitude of temporal resolution are expected

	Age range (BP)	
Resolution	General	Southwest
10^0 years	0–0.1	0–0.7
10^1 years	0.1–213	0.7–685
10^2 years	213–9425	685–21,605
10^3 years	9425–416,375	—

Note: The values are computed using the linear regression models discussed in the text.

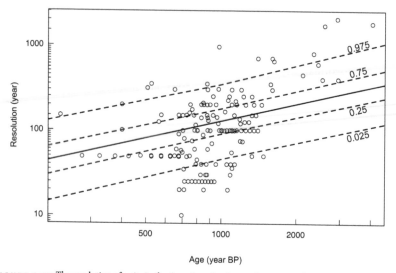

FIGURE 5.10: The resolution of units in the American Southwest decreases with time. The resolution of the units ($n = 197$) is plotted against their age on a log-log scale. The data points are not all independent measurements but are grouped by journal articles ($n = 30$).

The regression model predicts temporal resolution mostly on the order of centuries and millennia (table 5.3). An expected resolution in decades is found only in units that are younger than about 213 BP. Century-scaled resolutions are expected for units dating from 213 to 9425 BP. And millennial-scale resolutions are typical of units dating from 9425 to 416,375 BP.

Accounting for the spread of the data around the linear model, we expect 50% of the data points that are 100 years old to fall between 38 and 95 years and 95% of them to fall between 14 and 234 years. For 1000-year-old units, the 50% range is 159–397 years and the 95% range is 61–1004. By 10,000 years, the 50% and 95% ranges are, respectively, 628–1690 and 284–3572.

In the American Southwest, the expected resolution for a given age, on a log-log scale, is given by the linear model with an intercept of 0.11 and a slope of 0.66. The 95% central credible interval of the slope parameter is 0.45–0.78 (fig. 5.10).

The regression model predicts temporal resolution on the order of centuries for most of the region's prehistory (table 5.3). An expected resolution of decades is found in units that are younger than about 685 BP. Century-scaled resolutions are expected for units dating from 685 to more than 20,000 BP.

Accounting for the spread of the data around the linear model, we expect that 50% of the data points that are 100 years old will fall between 19 and 41 years and 95% of them will fall between 8 and 84 years. For 1000-year-old units, the 50% range is 90–179 years and the 95% range is 46–345 years.

Conclusion

The typical archaeological dataset contains units that are separated by hundreds of years and hundreds of kilometers and that lump together traces of human activities that may easily have taken place over centuries. Overwhelmingly, the sampling interval and the resolution of the data that appear in journal articles and in regional databases are greater than 1 human generation. In most cases, they are on the order of 10^2–10^3 years. Sampling interval and resolution of an order of magnitude of 10^1 years do exist, but they are rare. Spatially, at the hierarchical level of the site, the spatial sampling interval is on the order 10^1–10^3 kilometers.

The results highlight some of the points made in the two previous chapters. For instance, the nature of the human activities can play a significant role in shaping the quality of the archaeological record. In the data published in journal articles, the intervals of time between burials are typically shorter than those between nonburial contexts, suggesting that bioarchaeologists have access to data of higher quality than archaeologists do.

Similarly, the age of an archaeological deposit is an important determinant of its quality. The temporal sampling interval, spatial sampling interval, and the temporal resolution of archaeological units all degrade with time. While this is not surprising given the time dependence of many of the forces affecting the record, the analysis conducted here allows us to move beyond qualitative intuition and quantify, for the first time, the relationship between time and the quality of the record.

What is more, the different aspects of the quality of the archaeological record decrease with time following proportional rates that are smaller than 1. This means that the impact of age on the quality of the archaeological record diminishes with time, a result that is in line with other studies that have found taphonomic loss to decrease in magnitude over time (e.g., Surovell et al., 2009).

Several aspects of the quality of the archaeological record were not discussed in this chapter because the data to do so are either unavailable or difficult to collect systematically and on a large scale. This is the case for the spatial resolution. But it is safe to assume that the spatial resolution of archaeological contexts is probably better than their temporal resolution. For instance, discard in secondary refuse and disturbance processes are unlikely to move objects by more than a few dozens of meters, leading to assemblages with a spatial resolution of an order of 10^0–10^1 meters (Meltzer, 2004). But above the level of the site, analytical lumping is the dominant source of spatial mixing and can easily generate units with resolutions of hundreds if not thousands of kilometers.

More importantly, the dimensionality of the archaeological record has not been evaluated here. It would be difficult to measure in a systematic way the number of dimensions of datasets published in journals or in databases. But the background knowledge about how archaeologists analyze the record and about preservation loss tells us that archaeological data are highly dimensional with regard to certain phenomena and poor with regard to others. For instance, the archaeological record is highly dimensional when it comes to technologies. Archaeologists are trained to collect dozens of variables to describe objects like stone tools, ceramics, or basketry. Thus, when it comes to testing hypotheses about technological organization, archaeologists have access to dimensionally rich datasets. They can investigate, for instance, whether or not raw-material transportation distance increases rates of tool rejuvenation while controlling for covariates such as raw-material quality or the type of tool. The same dataset, however, is dimensionally poor when it comes to addressing questions that belong in the realm of behavior, society, or culture, such as foraging behavior or gender ideology, as most of the variables that shape these processes did not leave any direct material traces.

A lot of work remains to be done on evaluating, empirically, the quality of the archaeological record. Following the example of paleontologists, archaeologists should make the study of the quality of the archaeological record a component of their research program. Examples of questions that would fall under such a program include what is the typical sampling interval in different depositional environments (e.g., marine coast, temperate forest, alluvial plain)? How is sampling interval affected by the level of social complexity of the groups studied (e.g., hunter-gatherers, chiefdoms, states) or by the type of remains analyzed (e.g., lithic, ceramic, bone)? Only by treating the archaeological record as an empirical object, in and of itself, can archaeologists develop a sufficient understanding of the possibilities and the limitations of their data.

6

Archaeology and Underdetermination

Archaeologists pay lip service to the underdetermination problem. They know it exists but in practice act as if it does not. Archaeologists have repeatedly made claims about the past that, although consistent with the archaeological record, are not supported by any smoking gun. Indeed, it is difficult to find in the current literature an archaeological interpretation that does not demand some leap of faith from its reader. This is puzzling because when pushed, most archaeologists would readily admit that their data underdetermine much of their theory and that few of their claims are not based on conjectures.

This underdetermination problem originates from the discrepancy between the quality of the archaeological record and the way that the behavioral, cultural, and social explanatory processes used by archaeologists operate. In chapters 3 and 4, I reviewed the forces that shape the quality of the archaeological record. The number and the diversity of these forces strongly suggest that archaeologists should not hold their breath waiting for datasets with a quality similar to that of ethnographic data. This was confirmed in chapter 5, in which we saw that archaeological data are dominated by sampling intervals and resolutions of 10^2–10^3 years. What is more, very few dimensions of past human behavior, culture, and society are preserved in the archaeological record, allowing archaeologists to control for only a small handful of covariates. With a better understanding of the quality of archaeological data, we can now turn around and evaluate how archaeologists use this information.

The Processes Studied by Archaeologists Operate over Time Scales of <10^1 Years

Contemporary archaeology has two major goals: (1) the reconstruction of the cultural history of human populations and (2) the explanation of these cultural histories in terms of high-level causal processes (Binford, 1968a; O'Brien, Lyman, and Schiffer, 2005; Tolstoy, 2008; Willey and Sabloff, 1993).

The first thing that archaeologists do to achieve these two goals is to draw low-level inferences from the archaeological record. Low-level inferences are not so much about human behavior as about the nature of physical finds (Trigger, 1989). They include the identification of objects (e.g., this is a hearth; this is the distal part of the femur), typological classification (here is a Clovis point; there a sidescraper), and material identification (this is obsidian; this is iron).

Low-level inferences are not severely underdetermined by the archaeological record. First, being about the physical nature of objects, they naturally lend themselves to the smoking-gun approach, as the information required to make them tends to be abundant and nonambiguous in the archaeological record. Second, the number of hypotheses that compete with each other in the realm of low-level inferences is limited. For instance, only a very restricted number of competing processes can give rise to a charcoal stain surrounded by a ring of burned stones. Similarly, the list of lithic raw material available in a region is always limited and relatively short. For all these reasons, archaeologists rarely disagree on low-level inferences.

Using low-level inferences, archaeologists then turn to the "what," "when," and "where" questions of cultural history (i.e., the establishment of the sequence of events that marked human history) and interpretations in terms of invention, diffusion, inertia, migration, or trade (Lyman, O'Brien, and Dunnell, 1997; Tolstoy, 2008). Compared with low-level inferences, cultural historical processes interface more directly with human behavior and for that reason are more susceptible to being underdetermined by the archaeological record. For example, some of the tasks of cultural history, such as estimating the date of appearance and disappearance of traits, are vulnerable to the forces of loss that shape the archaeological record (chapter 4). And some processes, such as trade and migration, are often impossible to distinguish archaeologically.

But many of the basic questions of cultural history are amenable to the smoking-gun approach. As with low-level inferences, many cultural historical questions are very much about the physical world and material culture, which increases the chance that a smoking gun is found. And as is the case for low-level inferences, the number of hypotheses that compete in the arena of cultural history is often restricted. In fact, many cultural historical questions involve only two competing hypotheses. For example, was there or was there not a pre-Clovis occupation of North America? In addition, several cultural historical questions can be resolved without multidimensional data and the control of covariates. Instead, a single piece of physical evidence, such as a radiocarbon date, may be enough to constitute a smoking gun. This is why questions such as where sheep were first domesticated or when humans first reached Australia, while certainly not without controversy, often are, eventually, settled empirically and unequivocally.

But in the minds of most archaeologists, low-level inferences and cultural history are of secondary importance compared with the explanation of cultural histories. Indeed, the search for the causes of cultural change has defined the practice of North American archaeology since the 1960s. We train our graduate students for this task, and the majority of the literature we produce is presented as a contribution to our understanding of the processes underlying culture and society.

Yet, it is the task of explaining human history that is the most affected by underdetermination. The theories and processes archaeologists use for this task—those that are at the core of archaeology's research agenda—operate over short time scales of 10^1 years or less.

For instance, a growing number of archaeologists (myself included) have adopted cultural evolution theory as the main lens through which to interpret archaeological data. Cultural evolution theory (Boyd and Richerson, 1985; Cavalli-Sforza and Feldman, 1981) approaches culture as an inheritance system and describes the forces affecting cultural transmission. Many archaeologists posit that these forces are detectable in the archaeological record. But these forces operate quickly and can generate cultural change in an entire population in just a few years. Take, for instance, the force called "conformist-biased transmission." Conformist-biased transmission describes our tendency to conform to the majority and adopt the most frequent behavior in our group. Empirical studies of diffusion of innovations suggest that conformist-biased transmission can affect the frequency of cultural traits within a decade. For example, in the mid-1920s, a new hybrid type of corn seed diffused among farmers in Iowa (Ryan and Gross, 1943). Initially, the farmers were slow to adopt the new seeds. But the adoption rates eventually accelerated, peaking in 1936 and 1937, so that within 13 years most farmers were growing the new corn hybrid. The overall diffusion curve of this new corn seed (fig. 6.1) has two interesting features: it is S-shaped, and it is less than two decades long. The S-shape is a signature of conformist-biased transmission (Henrich, 2001), and the curve's duration indicates that conformism

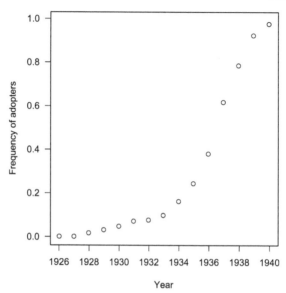

FIGURE 6.1: Diffusion of a new hybrid corn seed among Iowa farmers. The shape and the duration of the curve suggest that conformist-biased cultural transmission operates over decennial time scales. (Adapted from Ryan and Gross 1943; Henrich 2001.)

can lead to a population-wide shift in behavior in less than two decades. Other cases of cultural diffusion (E. Rogers, 1995) tell a similar story, suggesting, again, that the forces described by cultural evolution theory operate over time scales of 10^0–10^1 years.

The same is true for human behavioral ecology, a theoretical framework that is especially popular among archaeologists who study hunter-gatherers. Human behavioral ecology specifies, within the theory itself, the time scale over which the processes it describes operate. Behavioral ecologists start by making the assumption that natural selection has endowed individuals with the capacity to adjust their behaviors to changes in their social and ecological environment in ways that maximize their reproductive success. These behavioral adjustments are operationalized as decision rules. For instance, under a certain set of environmental conditions A, do x; under conditions B, do y. While the exact mechanisms by which individuals adjust their behavior optimally are black-boxed, the theory assumes that they are broad, flexible, and under minimal genetic, cultural, or cognitive constraints (E. Smith, 2000). As a result, humans can adjust their behavior to a wide range of environmental conditions and, more importantly, instantaneously. Behavioral ecologists thus expect human and other animal species to exhibit little adaptive lag, and they assume that variation in contemporary populations is mostly driven by variation in contemporary environments (E. Smith, 2000). Thus, behavioral ecological processes such as changes in diet, mobility regime, and mating patterns are predicted to take place very rapidly and well within an individual's lifetime (i.e., within time scales on the order of 10^1 years).

Similarly, agency theory describes processes that unfold over short time scales. The archaeologists who use agency theory are concerned with individual subjective perspectives and intentions as a source of change in the archaeological record. For instance, several ethnographic accounts of feasting suggest that feasting is used intentionally as a social and political tool to even out local and temporal variation in food supply, expand social networks, acquire prestige and status, or broadcast a costly signal (Hayden and Villeneuve, 2011). By definition, individual perspectives and intentions operate within the lifetime of individuals.

I did not cherry-pick these examples. Virtually every theory and process used by archaeologists describes mechanisms of change that operate over a decade or less. This is as true for the processual and evolutionary theories as it is for the humanistic ones. In fact, current archaeology has been marked by a renewed focus on processes that could be transposed to an ethnographic setting with no modifications whatsoever and that operate so fast that they could, theoretically, be observed by a cultural anthropologist within a single field season (Harrison-Buck, 2014; Kahn, 2013; Morehart, 2015). Similarly, when archaeologists were polled on what the most important scientific questions that the field will be facing over the next 25 years are (Kintigh et al., 2014), their responses focused on complex phenomena that unfold rapidly. Archaeologists are primarily interested in microscale processes.

Microscale processes have two related properties: (1) they operate within the span of a human lifetime (Dobzhansky, 1937) and usually within a decade; and (2) they operate at the hierarchical level of the individual or at a nearby level, such as the level of the household or the community, as is the case for processes like craft specialization, group identity, ideology, and power negotiation.

These microscale processes are used to interpret archaeological patterns that emerge at various scales ranging in order from 10^0 years and 10^0 km to 10^3 years and 10^3 km. For instance, an archaeologist may invoke microscale processes to explain why the activity areas at a site vary in content (10^0 years; 10^0 km) or to explain a temporal trend across three cultural time periods in a physiographic province (10^3 years; 10^3 km). This focus on patterns of 10^0–10^3 years is evidenced by the fact that the temporal scope of 86% of the journal articles surveyed in chapter 5 falls within the range of 10^1–10^3 years, with a median scope of 2174 years and a 25th–75th percentile range of 637–6201 years (fig. 6.2).

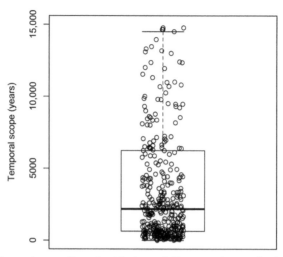

FIGURE 6.2: Temporal scope of journal articles (n = 402). The temporal scope of studies that appear in journal articles (see chapter 5) is calculated as the difference between the youngest and the oldest analytical unit to appear in an article. The data include datasets from general contexts, burial contexts, and the American Southwest.

Even the purported deep-time, macroscale archaeological studies, those that look at change over thousands of years and kilometers, boil down in the end to repeated rounds of microscale processes. This is what archaeologist Philip J. Arnold III (2008) calls "reel-time archaeology": the reconstruction of a series of ethnographic vignettes of the past, with the goal of viewing them in sequence, much like a movie reel, to obtain a lifelike animated picture of the past.

Most of these microscale processes are borrowed from source disciplines, such as cultural anthropology, psychology, and animal biology, that investigate individual behavior in the present time and have little concern for time and long-term trends. Archaeologists are transposing to the archaeological record a research program designed by, and for, scholars who have access to data quality that is better than that of the archaeological record by several orders of magnitude.

Not surprisingly, then, the vast majority of microscale processes are underdetermined by the archaeological record. First, most of them operate over time scales of 10^1 years or less. This is one, two, three, four, and sometimes five orders of magnitude faster than the sampling interval and the resolution of archaeological data (chapter 5).

Second, isolating the effect of microscale processes requires highly dimensional data. Humans are most complex at the hierarchical level of the individual—the very level at which most microscale processes operate. At an individual level, human behavior is little constrained. For instance, a phenomenon like religious belief can take a vast number of forms, whereas something like a projectile point cannot, constrained as it is by physical, mechanical, and functional factors. Thus, the set of competing hypotheses that can account for the social function of Upper Paleolithic cave paintings is vaster than the set of candidate explanations for why stone raw material shows signs of having been heated. The greater the degrees of freedom in a system, the more information is needed to study it, since it is more likely to have undergone significant changes within the intervals of time that separate observations or within the amount of time represented within the analytical unit, and since a greater number of independent processes can affect it. Indeed, a myriad of processes operate at an individual level, including decision making, personality, unconscious psychological biases, intention, social construction, age, gender, social interactions, norms, institutions, identity, life history, social learning, cultural evolution, political change, demographic change, weather, and seasons. Thus, the number of covariates that need to be controlled for to avoid false-positive or false-negative results is very large. What is more, most of these covariates do not directly involve material cultural and do not leave any traces whatsoever in the archaeological record. The test of consistency may allow archaeologists to turn a blind eye to this problem, but it does not make it disappear.

Finally, given the various forces that shape the quality of the record (chapters 3 and 4), the 100^{-3}-year and 100^{-3}-kilometer patterns that archaeologists seek to interpret may very well be false ones. For instance, the differential timing of adoption of agriculture between neighboring regions may be made consistent with several anthropological stories, but the timing could also be due to a Signor-Lipps effect (chapter 4). What is more, the information that would distinguish true from false patterns may have been irreversibly lost. In other words, the archaeological record can underdetermine the anthropological and taphonomic origins of its patterns.

Of course, there are exceptions—some microscale processes can be distinguished archaeologically. This is especially the case for the microscale processes that are close to the cultural history side of things. For instance, food preference and the trade of nonutilitarian items are individual-level, microscale processes. And yet, it is possible to infer that the people who occupied a region at a certain point in time preferentially consumed deer over other prey items or that they valued, somehow, the exotic birds that they imported from a distant region. But we cannot know why deer were preferentially selected or the cultural significance of the exotic birds. We may be able to make good guesses about the answers to these questions, guesses that are consistent with the data and the ethnographic record, but we will probably never have smoking guns to support them.

None of this implies that microscale processes are not responsible for the patterns that we see in the archaeological record. Quite the contrary: most processes that shape human material culture do so over short time scales of a decade or less. The very fact that rates of change in the archaeological record are dependent on the time intervals over which they are measured (see chapter 4) *implies* that material culture changes over time scales that are shorter than the sampling interval of the archaeological record: the processes that affect human material culture operate so fast that material culture has enough time to fluctuate back and forth or reach stasis within the interval of time that typically separates two archaeological samples. Archaeologists may thus be right that microscale processes are the main factors responsible for changes in material culture. But this does not mean that they can discriminate between these microscale processes, much as a detective may know that a crime occurred and yet be unable to identify the perpetrator.

As explained in chapter 1, rather than finding smoking guns that discriminate between competing hypotheses, archaeologists see it as their job to interpret their data through the lens of a theory in terms that are nothing more than consistent with the data. This has allowed a research program that is heavily underdetermined by the archaeological record to become mainstream. The gulf that separates the quality of the archaeological record and the research topics favored by archaeologists has three important consequences for the field.

CONSEQUENCE 1: MOST ARCHAEOLOGICAL RESULTS ARE WRONG

Inferences drawn despite an underdetermination problem are bound to be wrong. Until a smoking gun that can discriminate between all the plausible alternatives is found, there is no way whatsoever to know whether an interpretation is correct or not. And the greater the number of alternative explanations, the greater the likelihood of archaeologists' picking the wrong one—not wrong in the sense that "all statements in science are provisional and therefore bound to be wrong," but wrong in the sense that astrological predictions are bound to be wrong and, if they are right, are so purely because of luck. Astrology is still alive today because astrologers (and their clients) use the test of consistency, along with a generous serving of confirmatory bias, to turn a blind eye to the inconsistencies between astrological theory and the empirical world: the very same factors that have allowed the gap between archaeological theories and archaeological data to perpetuate and grow. This gap is so wide that archaeology today has more to do with pseudoscience than with science. If archaeologists were to apply the reasonable-doubt rule discussed in chapter 1, they would have to recant a large number of their claims about the human past. Indeed, how many archaeological results are supported by empirical data *beyond reasonable doubt*?

Would you want to live in a world in which detectives solved cases and put people in jail even when the evidence underdetermined the identity of the criminal? A world in which detectives start their investigation with one single hypothesis ("Colonel Mustard is the killer"), find evidence that can be made somehow consistent with that hypothesis (Colonel Mustard was in town the day the murder occurred), and imprison Colonel Mustard, without ever paying attention to the fact that other lines of evidence are equally consistent with other hypotheses (Professor Plum, Miss Scarlett, and Mister Green were also in town at the time of the murder)? And what about living in a world in which the safety of a new airplane design was determined with the same degree of confidence that we have in our interpretations of, say, the social role of clay figurines in Mesoamerica? Would you board that plane?

These rhetorical questions are not unfair. They do not set the bar too high for archaeologists. Archaeologists often lament that given how fragmentary the archaeological record is, it is unfair to hold them to the same epistemological standard adhered to by other disciplines. That if archaeologists were held to the same standard as physicists or biologists, they would never answer any interesting questions. And that we simply have to accept that archaeological inferences come with a high level of uncertainty. There are two flaws to these responses. First, the problem is not that we will not make any progress if we set the epistemological bar too high but rather that we will not if we set it too low. What matters is not whether we are doing the best we can with the data we have, or whether we are "at least trying" to explain cultural history in terms of high-level causes—what matters is that a leap of faith is a leap of faith and never a scientific result. It is always unwise to make up stories when we know that we lack the evidence to support them. Conjecture will always be part of the historical scientific process, but it does not have to loom so large as to threaten a discipline's entire epistemological status. That a smoking gun has not been found yet is not an excuse to fall back on the test of consistency and pick an interpretation. Second, these responses ignore the fact that underdetermination is relative not only to the quality of the data at hand but also to the type of questions that are being asked. The archaeological record, with its quality, underdetermines some processes more so than others. Nothing is forcing us to focus on the very processes that happen to be the most underdetermined by the record.

CONSEQUENCE 2: MUCH ARCHAEOLOGICAL RESEARCH IS UNNEEDED

By focusing on a research program that is underdetermined by the archaeological record, archaeologists have made themselves largely irrelevant to other disciplines. Even among university administrators, archaeology is perceived as a low-status discipline that has lost its relevance because it produces (granted, like many other social sciences) vast amounts of unverifiable information (Upham, 2004). But archaeological research is not irrelevant only because archaeological results are likely to be wrong but also because the research itself is unneeded. Archaeologists like to think of themselves as contributing to our understanding of the ahistorical processes that affect humans. Archaeologists do not merely borrow theories from various source disciplines, the thinking goes, but they also contribute to them. From the point of view of these source disciplines, however, archaeology has little to contribute. The psychologist who studies how craft skills are acquired by children does not need archaeology to confirm or disconfirm his ideas, and it is doubtful if he will ever learn anything new about craft skills acquisition from archaeologists using his theories to interpret the archaeological record. At best, he will treat the archaeological study-case as an interesting anecdote; at worst, he will dismiss it as pseudoscience, since studying skill transmission requires highly dimensional, high-resolution, fine-grained data that are difficult to obtain in contemporary contexts, let alone in the archaeological record. As Marion Smith ([1955] 1998, 173) pointed out more than 60 years ago, a large part of our research "may seem to an outsider, conscious of the weak logic involved [that of interpreting or explaining the past in ethnographic terms], that the subject has no sound intellectual basis at all." If what archaeologists are genuinely interested in are the processes described by their source disciplines, they should leave their trenches and their trowels behind and go study living populations. To borrow a question from paleontologists Stephen Jay Gould and Niles Eldredge (1977, 149), why be an archaeologist if we are to observe only very imperfectly what students of living populations can do directly?

CONSEQUENCE 3: ARCHAEOLOGICAL THEORY IS BALKANIZED

The underdetermination problem also causes archaeological theory to be heavily balkanized. The archaeological literature is crowded with a daunting number of theories and claims that are mutually exclusive (see Bentley, Maschner, and Chippindale, 2008; Hegmon, 2003; Hodder, 2001; Preucel,

1991; Preucel and Hodder, 1996; Upham, 2004). This is because new theories and processes are added to the literature faster than they are eliminated. Since the archaeological record underdetermines most processes that affect humans, the potential number of theories and processes that one can come up with and that can be made consistent with field observations is incredibly vast. What is more, the test of consistency shields these theories and processes from elimination. As a result, theories and processes accumulate quickly, disappearing only when they fall out of fashion.

This is also why debates about high-level archaeological explanations are rarely settled empirically. Rather than empirically testing competing hypotheses, different archaeologists interpret the archaeological record through the lens of their particular theoretical interests and thus settle on different explanations, even when they are looking at exactly the same data. It is thus not surprising that discipline-wide consensus for archaeological explanations—for instance, on the role that gender ideology played in patterning the material culture at Çatalhöyük—is exceedingly rare.

Why Archaeologists Ignore the Underdetermination Problem

How is it that archaeologists have come to pursue a research agenda that is underdetermined by their data? It is not because they are naive—archaeologists have long been aware of the mismatch between archaeological data and theory and the perils of uncritically borrowing theories from disciplines that study humans at the scale of individuals (e.g., Ascher, 1968; Bailey, 1981, 1983, 1987, 2007, 2008; Bar-Yosef and Van Peer, 2009; Dibble et al., 2016; Dunnell, 1984; Frankel, 1988; Garvey, 2018; Lyman, 2003, 2007; T. Murray, 1999; T. Murray and Walker, 1988; M. A. Smith, [1955] 1998; Stern, 1993, 1994).

As early as 1954, Christopher Hawkes, in a seminal paper, recognized that some aspects of prehistoric populations are more difficult than others to infer from the archaeological record—as illustrated by the so-called Hawkes's pyramid (fig. 6.3). Technologies, at the base of the pyramid, are the easiest to infer from the archaeological record. Above technologies, and thus more difficult to infer, are subsistence and economic systems, followed by sociopolitical institutions and, at the very top, religion and ideologies. Describing the difficulty of inferring sociopolitical institutions, Hawkes (1954, 161–62) notes that if "you excavate a settlement in which one hut is bigger than all the others, is it a chief's hut, so that you can infer chieftainship, or is it really a medicine lodge or a meeting hut for initiates, or a temple?" Hawkes also notes that as one climbs from the base to the top of the pyramid, one goes from processes that he calls "generically animal" to ones that are increasingly specific to humans.

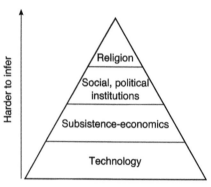

FIGURE 6.3: Hawkes's pyramid. (Adapted from Hawkes 1954.)

Hawkes comments that "the more human [a process is], the less intelligible [it is archaeologically]" (162); that is, as one moves up the pyramid, the less physically constrained processes are and the greater the number of competing explanations for the same piece of data. Hawkes's pyramid is a metaphor that captures the fact that the archaeological record underdetermines some processes to a greater extent than others. Processes that are higher up on the pyramid require higher-quality data to be studied. For instance, establishing that prehistoric shamans used access to long-distance trade networks to acquire psychoactive substances in order to justify their social status would require fine-grained and highly dimensional data that are unlikely to ever be found in the archaeological record, whereas determining what type of temper was used in the making of cooking vessels is perfectly feasible.

A year later, Marion Smith ([1955] 1998) followed up on Hawkes's critique in a paper that is very much in line with the argument laid out in this book. Although she never uses the terms "underdetermination" and "equifinality," Smith agrees with Hawkes that the archaeological record underdetermines several aspects of human behavior and culture. Not only that, but she also argues that claims that are underdetermined by archaeological data—however consistent they are with it—are not justifiable. Returning to Hawkes's example of a large hut, Smith says, "If you decide to call the large hut a chief's hut, and not a meeting house, or a temple, this is an assertion, not an argument. You can't really say that you know that it is, and if someone criticizes your assertion, it is impossible to provide sufficient evidence to convince him you are necessarily right. This is not the result of any fortuitous incompleteness in the archaeological record; the position couldn't be improved by better excavation, by finding a more favorable site, or by the invention of a new technique of analysis" (171). Given the quality of the archaeological record, Smith argues that "there are real and insuperable limits to what can legitimately be inferred from archaeological material" (172). For Smith, the archaeological record, by its very nature, is incompatible with goals such as re-creating the past or apprehending past societies. Because "unobtainable ends cannot be the proper ends for any subject," she says, we need to recalibrate our research aims "by strict reference to the potentialities of the evidence" (173).

Later, some archaeologists found inspiration in the French Annales school of history program, primarily the work of Fernand Braudel (1980). The historians of the Annales school point out that different historical processes operate at different temporal scales, ranging from individual historical events to the *longue durée*, a set of environmental and social structural constraints that can endure for centuries. Archaeologists have used Braudel's work to argue that archaeology also needs a hierarchy of explanation and that incompatibilities between socioeconomic processes and the archaeological record stem from interpreting the record at the wrong scale (e.g., Bintliff, 1991; Knapp, 1992; M. E. Smith, 1992). The Annales school, however, had a limited impact on archaeological thought, in part because the hierarchical levels it recognizes (events, conjectures, *longue durée*) are somewhat rigid and poorly defined and because, although it sought to identify processes that unfold over long time spans, it did not speak directly to how assemblages that formed from the accumulation of processes operating at various time scales should be interpreted (Holdaway, 2008). As a result, its influence mostly extended to archaeologists who work on recent time periods and who deal with a quality of data that is similar to that of historians (Bailey, 2007; Fletcher, 1992).

In a string of papers published over three decades, Geoff Bailey spearheaded "time perspectivism," an approach that also emphasizes the idea that different processes are observable at different time scales (Bailey, 1981, 1983, 1987, 2007, 2008). Time perspectivism rests on two pillars: (1) All archaeological artifacts are palimpsests; that is, they are time and space averaged. (2) These palimpsests are both a blessing and a curse. They are a curse because they make the processes described in contemporary social theory difficult to infer archaeologically and render suspicious the legitimacy of many popular archaeological concerns, such as finding tool kits, activity areas and living floors, assigning function to clusters of artifacts and features, and estimating population size (Wandsnider, 2008).

But palimpsests are also a blessing, time perspectivism contends, as they can bring into focus processes that are not visible over fine time scales, an idea also put forward independently by both Lewis Binford and Robert Foley the same year that Bailey published his first paper on time perspectivism (Binford, 1981; Foley, 1981). More than a mere problem that needs to be corrected, palimpsests offer the opportunity to observe long-term processes. Time perspectivism, like the French Annales school of history, calls for scale-dependent explanations: archaeologists ought to focus on those processes that unfold over time scales that are commensurate with the resolution of archaeological data.

As an approach, time perspectivism never gained significant ground (Bailey, 2008). From England, where it originates, it spread primarily to Australia and New Zealand, where archaeologists such as Tim Murray and Nicola Stern took Bailey's ideas and refined them significantly, emphasizing the links between site formation processes and time averaging (Fletcher, 1992; Frankel, 1988; Holdaway, 2008; T. Murray, 1999; T. Murray and Walker, 1988; Stern, 1993, 1994, 2008). Developments in the study of site formation processes, artifact reuse, and tool life history over the last 30 years have led to a wider adoption of time perspectivism, including by North American archaeologists, as illustrated by the contributions in Holdaway and Wandsnider 2008. Whereas these empirical studies have been successful at confirming that palimpsests are a universal phenomenon (see also Olivier, 2011), they fell short when it came to extracting compelling and substantive examples of long-term processes from the archaeological record (Bailey, 2008).

There are multiple reasons why time perspectivism never became as popular as it deserves. Some of these reasons are reviewed by Bailey himself in a retrospective of time perspectivism (Bailey, 2008) and are similar to the reasons why archaeologists ignore the general problem of underdetermination: an unshakable belief in the primal importance of the individual as an object of study and in the idea that the output of archaeology ought to take the form of a story with a narrative structure. Time perspectivism also suffered from a lack of clarity on how it is to be implemented (Bailey, 2008). The approach argues for the search for long-term processes in the "varying resolutions of different palimpsests" (Bailey, 2008, 26), but what these long-term processes might be is never made clear. Time perspectivism lacks a methodological précis, even if only the sketch of one, that would guide archaeologists in their quest for long-term processes. This lacuna stems in part from a reliance on fuzzy terms, such as "scale" and "temporality," with meanings that shift from one author to another.

But more importantly, perhaps, time perspectivism is overwhelmingly concerned with only one aspect of the quality of archaeological data: its temporal resolution. This near-exclusive focus on temporal resolution explains why many time-perspectivist studies have focused on intrasite analyses of palimpsests—a spatial scope of observation that is likely too narrow to avoid the underdetermination problem and unravel any compelling large-scale processes. In many ways, this book is an extension of time perspectivism: in addition to searching for "scale-dependent" processes (i.e., resolution-dependent ones), I argue that we also need to search for scope-, sampling-interval-, and richness-dependent processes.

The critiques of archaeology's research program—by time perspectivism, the Annales school, by Marion Smith, Christopher Hawkes, and many others over the last four decades (e.g., Barton and Riel-Salvatore, 2014; Binford, 2001; Clarke, 1968; deBoer, 1983; Dibble et al., 2016; Dunnell, 1982, 1984; Frankel, 1988; Garvey, 2018; Holdaway and Wandsnider, 2006; Lucas, 2005, 2012; Lyman, 2003, 2007; Meltzer, 2004; Olivier, 2011; Ramenofsky, 1998; Shennan, 2002; M. E. Smith, 1992; Vaquero, 2008)—have been largely ignored or perceived as misplaced pessimism and a lack of resolve at improving the quality of archaeological data. This extends even to Paleolithic archaeologists, who contend with a record whose low quality is definitely impossible to ignore, and yet who still try to map processes about hunter-gatherer mobility and social organization that operate over days or seasons on assemblages that accumulated over centuries (S. Kuhn and Clark, 2015).

Archaeologists ignore the underdetermination problem mainly because of (1) the history of their discipline, which has led them to become interested in ethnographic-scale processes that are not commensurate with the archaeological record. This problem is further amplified by (2) an incomplete definition of uniformitarianism, (3) the psychological pull of a human-centric view of the world, and (4) the way archaeologists are trained, especially with regard to hypothesis testing and the narrow way they think about the quality of their data.

FACTOR 1: THE HISTORY OF THE DISCIPLINE

In America, the failure of archaeologists to recognize the underdetermination problem can be traced back to a mistake made in the early days of the field. The first practitioners of American archaeology believed that the archaeological record of the New World was very recent. They thought so largely because they lacked chronometric dating techniques and because the European model of stone tool evolution, which suggests that the prehistoric cultures of Europe have changed significantly over time, did not seem to apply to the archaeological record of America (Lyman, 2007; Meltzer, 1985, 2005; Trigger, 1989; Willey and Sabloff, 1993). Rather, anthropologists like Franz Boas (1902) and Alfred Kroeber (1909) saw no significant differences between the material culture of prehistoric populations and contemporaneous Native American tribes. As a result, they assumed, mistakenly, that the American archaeological record had been produced by cultures similar to the ones observed ethnographically, and thus they came to view archaeology as prehistoric ethnology (Lyman, 2007), setting, by the same token, the discipline on the path it is still on today.

By going down the "prehistoric ethnology" path, archaeologists sought to interpret the macroscaled record in terms of microscale anthropological theory. Archaeology became a subordinate to cultural anthropology. As early as the 1940s, Walter Taylor described archaeologists as paltry technicians whose job is to recover cultural information that is to be interpreted by "those who have made it their business to study culture, namely anthropologists" (Taylor, 1948, 43). From the outset, American archaeology stood not only as a subfield of anthropology but as a minor one at that.

The idea that archaeologists must explain archaeological data through the lens of cultural anthropological theory was magnified by the advent of processual archaeology in the 1960s (Lyman, 2007). Processual archaeologists, under the mantra "archaeology is anthropology or nothing," endeavored to elevate archaeology above the rank of "history" and to the rank of "science" (Binford, 1962, 1964, 1965, 1968a). Lewis Binford, the most prominent architect of the processual agenda, though he would later revise his view, promoted the idea that the archaeological record is nothing less than "fossilized human behavior" (Binford, 1964, 425) and that most, if not all, components of prehistoric cultural systems are preserved, directly or indirectly, in the archaeological record (Binford, 1962, 1964, 1968a). Binford (1962) argued that archaeologists can, and should, contribute to cultural anthropology by studying these past cultural systems.

In Europe, a similar change in the goals of archaeology was propelled by the radiocarbon revolution. The advent of radiocarbon dating, by showing that Europe's megaliths were older than Egypt's pyramids and the temples of Malta, forced archaeologists to find other explanations for Europe's megaliths besides their diffusion from the Near East (Renfrew, 1973). With the diffusion paradigm shattered, European archaeologists working during the end of the 1960s shifted from "talk of artifacts to talk of societies" (Renfrew, 1973, 253). The new theoretical framework that developed during these years emphasized understanding the societies behind the artifacts recovered by archaeologists. Practically, this meant explaining the changes observed in the archaeological record in terms of population growth, population density, and subsistence patterns, and how these are linked to changes in social organization, beliefs, art, and religion (Renfrew, 1973).

The processual archaeologists of the 1960s, in America and Europe, did recognize the incomplete nature of the archaeological record, but they saw it as a methodological problem, a technicality that needed to be dealt with, rather than as an epistemological barrier. The processual archaeologists reasoned that before archaeologists can study these past cultural systems, they must translate the archaeological record into an ethnographic record. This task is the domain of the so-called "middle-range theories," the theories that describe the functional link between the various aspects of human behavior and material culture (Binford, 1968a), and whose purpose is to bridge the static archaeological record and the dynamics of past human behaviors that created that record (Binford, 1977). If only archaeologists are clever enough in building these middle-range theories, Binford (1968a, 23) argued, they can reconstitute prehistoric cultures in their entirety: the "limitations on our knowledge of the past are not inherent in the archaeological record; the limitations lie in our methodological naivete." It is up to us to find a way to infer, say, a prehistoric group's kinship system from its material culture (Binford, 1962). Archaeologists thus started conducting systematic ethnoarchaeological and experimental studies in an effort to build these middle-range theories.

This view of archaeology as ethnography of the past led archaeologists to mine cultural anthropology and other disciplines for theories and processes, which further amplified the underdetermination problem. Indeed, most of these source disciplines study processes that operate over short time scales that are rarely longer than a few years. Most of them have short scope, either because they are relatively new disciplines (e.g., primatologists started collecting observations about primate behavior only in the 1950s) or because the human life span is limited: individual scientists do not get to conduct active research for more than a few decades, and ethnographers are unlikely to observe more than four human generations in their fieldwork. Like blinders on a horse, these short scopes limit the range of processes they can observe to the fast-acting ones (see chapter 2).

Starting in the late 1960s and the 1970s, research in ethnoarchaeology and experimental archaeology raised awareness about how the archaeological record forms (O'Brien, Lyman, and Schiffer, 2005), and archaeologists started to question the premise that the archaeological record is a direct reflection of behavioral processes. This questioning included Binford himself, who, in a change of mind, argued in 1981 that the archaeological record is a palimpsest that is irremediably different from the dynamic record. Binford, along with others (Bailey, 1981; Foley, 1981), saw the palimpsest nature of the record as offering an opportunity to focus on human behavior at a hierarchical scale that is inaccessible to ethnographers. But this realization did little to change the general aims of archaeology, and site formation processes remain an unfortunate methodological hurdle that complicates archaeological research but that, with careful investigation, can be controlled for in order to reveal small-scale behavioral events (Reid, Schiffer, and Rathje, 1975; Schiffer, 1976, 1987).

To this day, the view that pervades archaeological thought is that archaeological assemblages need only to be cleared of the distorting effects of site formation processes and fleshed in with ethnoarchaeology or experimental archaeology before they can be used as proxies of actual observations of individual behaviors (Holdaway, 2008). Indeed, the literature is filled with attempts at isolating "living floors" and discrete synchronic behavioral episodes using "ethnographic excavation" methods such as *décapage* (see Dibble et al., 2016, for a list of examples).

And to this day, the principal source of archaeological research questions and interpretations remains anthropological theory, the bulk of which is based on observations made in the ethnographic present (Lyman, 2007). However we may view the processualist agenda—as naive, incomplete, or simply wrongheaded—it cemented the view that archaeologists should interpret their data in the frame of reference of cultural anthropology or of other fields that study humans in contemporary settings. This view is evidenced by the current trend for an individual-centered archaeology, which is marked by a concern for the subjective experience of individuals, their intentionality, their creativity, and their actions (e.g., Gamble, 1999; Gamble and Porr, 2005; Hodder, 2000; M. L. Smith, 2010), and by the fact that the reconstruction of past lifeways is still listed as one of the objectives of archaeology in textbooks and classes the world over (Holdaway, 2008).

FACTOR 2: AN INCOMPLETE DEFINITION OF UNIFORMITARIANISM

A widespread incomplete definition of the principle of uniformitarianism also contributes to the view of archaeology as ethnography of the past. Uniformitarianism refers to the assumption that there are universal principles that apply irrespective of time and space (S. Gould, 1965). But many archaeologists take uniformitarianism to mean that the archaeological record is the product of the same processes that are observable in the present (the same mistake was made by paleontologists; see S. Gould, 1965; Shea, 1982). This definition of uniformitarianism is not wrong per se, but it is incomplete and misleading. It is easy to conclude from it, mistakenly, that the archaeological record ought to be explained in terms of the processes that are visible on a daily basis in the ethnographic present (Bailey, 1981, 1983). This conclusion fails to acknowledge the fact that different processes can act uniformly in time and space but at different rates. Not all the uniformitarian principles that affect human behavior (if they exist at all) have to be observable in the present. Some uniformitarian principles may be operating too slowly, or over too wide a spatial scale, to be detectable ethnographically (Bailey, 1981, 1983). The incomplete view of uniformitarianism further compels archaeologists to force their data into the frames of references of cultural anthropologists and other source disciplines concerned with the study of human behavior in the present. Equally critical, it also draws archaeologists away from studying slow-acting, large-scale processes that cannot be reduced to short-scale ones.

FACTOR 3: AN ANTHROPIC BIAS

Archaeologists are also pulled toward short-term processes because of an anthropic bias. We all view and experience the world at the scale of a human individual. It is natural for us to seek to explain the world in terms of those processes that are the most familiar to us, that is, those that operate over time scales shorter than our lives and over spatial scales smaller than those that our social groups occupy. Naturally, this leads to the still strongly held belief that the individual is the most, if not the only, relevant object of study (Bailey, 2008). This pull of the familiar also makes it difficult for us to wrap our heads around vast expanses of space and time. Processes that operate over millennial or global scales are hard to grasp intuitively, feel foreign, and, as a result, are less likely to become objects of scientific inquiry. The scale of the archaeological record, in particular, is difficult to comprehend, and archaeologists have hardly been able to resist the psychological tendency to couch their research, whether it is about a particular place, people, or even global history, in the form of a linear story that unfolds in a way that is similar to how the individual experiences time and is shaped by the same processes that we can observe in our daily lives (Bailey, 1981, 2008; Clarke, 1968).

FACTOR 4: THE WAY ARCHAEOLOGISTS ARE TRAINED

Finally, the way archaeologists are trained shields them from having to recognize the underdetermination problem.

First, most archaeologists have been taught, perhaps implicitly, to confirm hypotheses instead of thinking of fieldwork as a quest for a smoking gun. In fact, many archaeologists think of their research not as a hypothesis-driven enterprise but rather as an interpretive one. The test of consistency allows them to explain their data in terms of their favorite theory while ignoring the fact that their interpretation is only one of a large set of equally consistent explanations (Bailey, 1981, 1983; Stern, 1993).

What is more, the test of consistency makes archaeologists prone to the confirmatory bias and causes them to be deeply committed to their theoretical perspectives (chapter 1). Think how rare it is for a finding to ever have the power to overturn one's theoretical orientation (T. Murray and Walker, 1988). It is even more difficult to resist confirmatory-biased thinking when one has been trained in an environment in which mentors, reviewers, editors, and peers all participate in interpretive, confirmatory research.

Second, archaeologists are not trained to think clearly about the quality of their data. And they are not trained to consider all the different pathways by which underdetermination can creep into their research programs. Whereas archaeologists all know that their data are incomplete and distorted in some ways, they tend to focus on one aspect of data quality instead of considering the data's scope, sampling interval, resolution, and dimensionality all together.

A focus on resolution, for instance, may lead archaeologists to overestimate the number of processes that they can study and to fall back into the underdetermination trap. An archaeologist may mistakenly conclude that she can study a process that has been observed in the ethnographic record because her site contains a sequence of fine-grained stratigraphic units. Or an archaeologist may argue that because small-scale events representing hours, or even minutes, of activity time, such as a burial, the knapping of a stone implement, or the butchery of an animal, are visible in the archaeological record, they can be studied (Lucas, 2005). Along the same lines, one of the ways in which Paleolithic archaeologists tend to respond to the mismatch between their theories and the quality of their data (when they respond at all) is to concentrate on those exceptional locations with "ethnographic-scale" resolutions (S. Kuhn and Clark, 2015) and so-called "living floors" (Dibble et al., 2016).

The problem with this is that exceptionally fine-grained resolutions are, by definition, exceptionally rare and form a poor sample of the archaeological record. More importantly, visibility does not mean studiability. Isolating the causes behind an outcome requires more than observing the outcome. Fine resolutions are necessary but not sufficient to study short-scale processes. Even a series of cultural layers with extraordinarily fine resolution will, in all likelihood, underdetermine short-scale behavioral processes, be it because the intervals between them are too large or because the layers do not have the dimensionality to control for the relevant covariates. What is more, the archaeological record can underdetermine its own resolution: a site or a context that appears to represent an ethnographic snapshot may, in reality, have accumulated over decades or centuries. Because of the forces of mixing and loss, the possibility of false positives (e.g., two objects are interpreted as belonging to the same small-scale event when they do not) or false negatives (e.g., evidence that should have been there is missing) is always looming, even in fine-grained contexts.

Another response of Paleolithic archaeologists to the mismatch between theory and data is to accept the coarse-grained nature of the record and interpret it in terms of repeated rounds of short-scale behavioral processes (S. Kuhn and Clark, 2015). Positing that the variation between coarse assemblages is due to repeated rounds of some short-scale behavioral process does nothing to solve the problem of a mismatch between theory and data. If two processes are equifinal over short time scales, chances are that their long-term effects are also equifinal. And even if two processes do lead to different outcomes over the long term (perhaps because they operate at different rates), coarser resolution also increases the chance that a third, a fourth, a fifth, or many more additional and equifinal processes also operated and shaped the data at hand.

Archaeologists also have minimal exposure to studies of large-scale processes. Over the course of their careers, archaeologists see hundreds of examples of archaeological cases involving microscale, individual-level processes but seldom any cases of truly large-scale ones. This makes it difficult for archaeologists to even conceive of what large-scale processes they could possibly study (Bailey, 2008). In addition, few efforts have been made to measure empirically and quantitatively the different aspects of the quality of the archaeological record. This, in turn, makes it difficult to delineate the range of research questions that can be answered properly. Understanding what quality of data we can reasonably expect from the archaeological record, and taking into account not just the resolution of our data but also its scope, its sampling interval, and its richness, allow us to narrow down more rigorously, like so many circles in a Venn diagram, the range of processes that archaeologists ought to study. This, in turn, helps us define a more practical implementation program for the search for long-term phenomena.

Finally, the archaeological community inhibits the recognition of the underdetermination problem. Even if a young archaeologist were to free herself from the blinders of her training and recognize the underdetermination problem in her research, there is little incentive for her to modify her research agenda accordingly. A historical scientist facing data that underdetermine their cause must remain agnostic about what that cause may be, and agnosticism does not make for catchy papers. By insisting on remaining agnostic about a past phenomenon, our young archaeologist would swim against the tide of what the great British biologist C. H. Waddington (1977) calls the "Conventional Wisdom of the Dominant Group." She would frustrate her thesis committee. She would publish fewer papers and in less prestigious journals.

She would have a harder time finding an academic job and getting tenure. Changing one's research program to avoid an underdetermination problem is a frequency-dependent strategy: when rare, it is a high-cost, low-benefit strategy. This is even truer when funding agencies favor "useful" research that provides solutions to contemporary societal problems, nudging archaeologists to link the archaeological record to events that are observed in the present.

Paleontology Overcame the Same Underdetermination Problem

Like the proverbial frog in hot water, archaeologists have grown accustomed to the underdetermination problem and to claims about the past that are based on conjectures and leaps of faith. But the history of another field, paleontology, tells us that it does not have to be that way. Like archaeologists, paleontologists suffered an underdetermination problem that arose from a mismatch between the quality of the fossil record and the theories they used to interpret that record. But unlike archaeologists, they overcame their underdetermination problem during the 1970s, as their field underwent an epistemological revolution (D. Sepkoski, 2005, 2012; J. Smith, 1984).

Before this epistemological revolution, paleontology played a subordinate role in biology. Like archaeologists, paleontologists had been trying to make sense of the fossil record in terms of a microscale theory—in their case, evolutionary biology. Ever since the modern synthesis of the 1940s, evolutionary biology has been emphasizing changes in gene frequency as the main process of biological evolution. Because of this, geneticists and other laboratory biologists became the main players in the field. And since paleontologists cannot recover direct evidence of genetic transmission in the fossil record, they were relegated to the rank of cheerleaders, whose job was to discuss how the fossil record is consistent with microevolutionary theory (J. Smith, 1984; Valentine, 2009).

At the time, paleontologists were conducting two types of research. The first type resembled archaeology's cultural history; it consisted in reconstructing the history of life on earth by mapping the spatial and temporal distribution of fossil groups and phenotypic traits. The second type of research was similar to archaeology's explanation of history and consisted in reconstructing the behavior and biology of fossil organisms to create something that a field biologist would recognize—living organisms (Turner, 2009, 2011). What were the members of this taxon eating? Were they nocturnal or diurnal? Sociable or solitary? The different features of these fossil organisms, and how they changed through time and space, were explained in terms of evolutionary biology theory.

With this research agenda, paleontologists made themselves unneeded by biologists, playing only a "marginal role compiling a photo album of the history of life on earth" (Princehouse, 2003, 6). Their reconstruction of the history of life was deemed useful by the other biological disciplines, but their explanation of this history in terms of microevolutionary theory was not. The processes that paleontologists claimed to observe in the fossil record actually demanded a data quality that the fossil record could simply not provide. Most of the processes described by evolutionary biology theory are better observed among living organisms or in test tubes. For example, population geneticists do not need paleontology to confirm the existence of a process like the founder effect or to better understand how it works. But paleontologists' confirmations of these microevolutionary processes were more than unneeded: they were also unconvincing, as the fossil record underdetermines microevolutionary processes. In addition, by confining themselves to an interpretive and confirmatory agenda, paleontologists could not make theoretical contributions and propose novel evolutionary mechanisms of their own (S. Gould, 1980; J. Smith, 1984). As John Maynard Smith (1984, 402) put it, "the attitude of population geneticists to any paleontologist rash enough to offer a contribution to evolutionary theory has been to tell him to go away and find another fossil, and not to bother the grownups."

During the 1970s, however, the agenda of the discipline was transformed, not only by the adoption of quantitative tools such as mathematics, simulation, and statistics but also, and more importantly, by a new way of reading the fossil record (S. Gould, 1980; Jablonski, 1999; D. Sepkoski, 2005, 2012). Paleontologists began to look at the fossil record as a record of macroevolution rather than of microevolution. In biology, the term "microevolution" denotes the processes that are observable within the span of a human lifetime, while "macroevolution" refers to those that are observable on geological time scales (Dobzhansky, 1937). What the paleontologists who led the epistemic revolution did was to recognize that the quality of the fossil record is inadequate to study microevolution and that expecting the fossil record to measure up to a programmatic agenda designed by, and for, microevolutionists would always remain unproductive. The only viable solution was to recalibrate their research interests to the quality of the fossil record. In this recalibrated research program, the reconstruction of the biology and the behavior of past organisms was largely replaced by a search for macroevolutionary processes that operate above the species level, such as extinction rates and species selection (Turner, 2009).

You may be surprised by how far this revolution went. Henry Gee, a British paleontologist and senior editor at *Nature*, notes that some paleontologists in the 1970s gave up on trying to understand the adaptive function of traits (why did birds start to fly?) or on trying to place fossils in a sequence of ancestry and descent (did *Homo sapiens* evolve from *Homo erectus*?)—the kind of questions that had been the bread and butter of the field for most of the century (Gee, 1999; see also R. Smith and Wood, 2017, for a paleoanthropological treatment of similar ideas). Given the quality of fossil data, Gee notes, any adaptive scenario that links birds to flight is elusive and will remain so for a long time, if not forever. This is not to say that natural selection is not an important force that has shaped life on earth. Rather, it is to say that it is impossible to identify in the fossil record the particular selective pressures that led, say, some fish to grow legs.

Consider how hard it is to observe natural selection even in contemporary organisms. While adaptations are ubiquitous in nature, the cases where natural selection has been seen in action in the wild are rare. Most often, biologists can only collect data that suggest the presence of natural selection (Endler, 1986). This is because several conditions have to be met in order to isolate the effect of natural selection from other processes, and these conditions are hard to meet. To demonstrate the action of natural selection, a researcher needs to show that Darwin's three conditions for natural selection have been met. First, the researcher needs to show that there is a superabundance of individuals—that the environment cannot support the entire population and that not every individual can reproduce with the same success (Darwin's first condition). The researcher also needs to document that phenotypic variation among individuals exists and that it translates into variation in the individual's ability to survive and reproduce (Darwin's second condition). Finally, this phenotypic variation must be shown to be heritable (Darwin's third condition). After showing that these three conditions are met, the researcher needs to identify the source of the selective pressure and rule out alternative sources of phenotypic variation, such as migration, all of which requires detailed quantitative measurements of the environment and the study population— over multiple generations.

The most famous and compelling case of natural selection in the wild comes from a study of Darwin's finches on Daphne Major, a Galapagos Island, by Peter Grant and Rosemary Grant (1986). The insularity of Daphne Major allowed the Grants to rule out migration as a source of phenotypic change in their study population. What is more, the island possesses a simple ecosystem, which facilitated the identification of the selective pressure that acted upon the finch population: fluctuation in the availability of different kinds of seeds caused by yearly changes in environmental conditions. What is more, the population of Darwin's finches on the island is small, so that it was feasible for the Grants to collect precise quantitative data about the beak morphology and the diet of most of the individuals in the population, over 20 consecutive generations. The Grants demonstrated that there is competition for resources, especially during the dry season, when food is scarce and dominated by large, tough seeds (first condition). They also showed how variation in beak morphology causes variation in the birds' capacity to process different kinds of seeds and, thus, to variation in survival (second condition). Finally, they were able to show that beak morphology is heritable (third condition). Thus, in order to demonstrate that the beak morphology of Darwin's finches is an adaptation to different kinds of seeds, the Grants collected multidimensional data (among the dimensions measured: the beak morphology of individual birds, their diet, their number of offspring, the availability of different plant foods, the characteristics of different seeds) with a short sampling interval (data points are separated by just a few months) and with high resolution (the data were collected at a seasonal scale, shorter than one finch generation). It would be impossible to obtain this kind of information in the fossil record, which is why the fossil record underdetermines natural selection and, for that matter, any other microevolutionary process.

The recalibration of paleontology's research agenda can be interpreted as an attempt to reduce the underdetermination problem by focusing on the processes that can be observed in the fossil record, given its scope, its sampling interval, its resolution, and its dimensionality. It is thus not surprising that the recalibration was accompanied by an effort to better understand the quality of the fossil record.

Paleontologists made the incompleteness of the fossil record an object of study in itself (Turner, 2011), and discussions about the quality of the fossil record have occupied a prominent place in the paleontological literature ever since (e.g., Behrensmeyer and Chapman, 1993; Behrensmeyer, Kidwell, and Gastaldo, 2000; Fürsich and Aberhan, 1990; Jackson and Erwin, 2006; Kidwell and Bosence, 1991; Kidwell and Flessa, 1996; Kidwell and Holland, 2002; Kowalewski, 1996; Kowalewski and Bambach, 2003; Kowalewski, Goodfriend, and Flessa, 1998; Martin, 1999; Raup, 1979; Walker and Bambach, 1971).

Focusing on those processes that are actually observable in the fossil record, that are not merely the products of repeated rounds of microevolution (Erwin, 2000; S. Gould, 1980; Jablonski, 1999; Turner, 2011), and that cannot be studied by biologists in the laboratory or in the field allowed paleontology to become a source of challenge and modification to evolutionary theory (Princehouse, 2003) and, eventually, to be welcomed to the high table of evolutionary biology (J. Smith, 1984).

Archaeologists can take inspiration from the success of the paleontological revolution. This does not mean that archaeologists should directly transpose paleontology's agenda to the archaeological record, nor is it that the fossil record and archaeological record are the same and that archaeologists should view artifacts as species. Rather, it means that, like paleontologists, archaeologists can solve the mismatch between the quality of their data and their research program by changing their research program. This is how archaeologists will take full advantage of the archaeological record and its quality.

7

Taking Advantage of the Archaeological Record

Despite more than 60 years of exhortations (e.g., Bailey, 1981, 1983, 1987, 2007, 2008; Barton and Riel-Salvatore, 2014; Binford, 2001; Dibble et al., 2016; Dunnell, 1982, 1984; Fletcher, 1992; Frankel, 1988; Garvey, 2018; Hawkes, 1954; Holdaway and Wandsnider, 2008; Holdaway, 2008; Holdaway and Wandsnider, 2006; Lyman, 2003, 2007; Meltzer, 2004; T. Murray, 1999; T. Murray and Walker, 1988; Shennan, 2002; M. A. Smith, [1955] 1998; M. E. Smith 1992; Stern, 1993, 1994; Vaquero, 2008), archaeologists are still primarily interested in microscale processes that unfold over time scales of decades or less. Given the quality of the archaeological record, archaeologists could not have picked worse processes to study.

By emphasizing microscale processes, archaeologists are not only misusing the archaeological record but also underusing it. Indeed, archaeologists have yet to take full advantage of the archaeological record and its contributive value to the social sciences (beyond the contribution of cultural history). To do that, archaeologists need to recalibrate their agenda to the quality of the archaeological record. This recalibrated research program is very different from the one that currently defines the field. It evacuates the study of most individual-level processes and prioritizes instead two tasks: (1) the reconstruction of cultural histories and (2) the search for macroscale patterns and processes in the global archaeological record.

Recalibrating archaeology's research agenda is the only viable solution to the underdetermination problem that plagues archaeology. Whereas technological breakthroughs and other methodological advances will continue to expand the amount of information archaeologists can extract from the record, these gains will never change fundamentally its expected quality. At the mercy of nature, archaeologists have to contend with the fact that there is a strictly limited amount of information about the human past that has been preserved on the earth's surface. This amount of information, and its quality, set boundaries on the range of processes that can be studied. It is up to the archaeologists to work within those boundaries or not.

The field's agenda should also be adjusted for ethical reasons. The archaeological record is a finite resource, much like petrol. Segments of the record already appear to be near depletion, as rates of discovery in a broad variety of sites, such as Classic Maya monumental centers, European Upper Paleolithic sites, proboscidean kill sites, and shipwrecks, are in decline (Surovell et al., 2017). Before archaeologists go out to the field and destroy another portion of the record, they should ensure that it is to answer a question that is (1) answerable given the quality of the archaeological record and (2) unanswerable given the data that have already been collected.

It falls on each and every archaeologist to evaluate, on a case-by-case basis, the match between the quality of their data and the phenomenon they are interested in. Can I answer this research question *beyond reasonable doubt?* Can I show that the traces recovered from the field are consistent with this particular cause and this cause only? These questions are not easy to answer. First, archaeologists need to think clearly about what hierarchical level they need to work at, since some processes are observable at some levels and not at others (chapter 2). For instance, the impact of conscription on gender division of household labor will be best observed with household-level data. Then, archaeologists need to estimate the minimum and maximum scale over which the process of interest can affect material culture. How quickly can this process affect material culture? In the case of conscription and household labor, the answer is probably days. And what would be the longest a process takes to affect material culture? At the most, conscription would take a year to affect household division of labor, such as when there is a seasonal cycle to household labor. Similarly, the minimum and maximum spatial scale over which the phenomenon is expected to affect material culture needs to be estimated. The answer to all these questions will determine the scope, the sampling interval, and the resolution scale that are needed to study the process (see fig. 2.5). Finally, archaeologists need to build a list of competing explanations for the pattern observed. For example, what, besides conscription, could generate a lack of evidence for male-related craft production activities in households occupied during a period of warfare? An absence of male-related domestic craft production could be due to the fact that the assemblage analyzed happens to represent a season during which men do not participate in household activities. Or it could represent a season during which men's and women's contributions to craft production are the same. Another possible explanation is that the war affected the economy and resulted in a dwindling demand for the goods produced by men. The impact of the war on trade networks may also have affected the availability of raw material necessary to produce the men's craft goods. Or women may have taken over the production of craft goods for cultural or social reasons that have nothing to do with conscription. The list of competing explanations will determine what covariates need to be measured in order to discriminate between the competing hypotheses and shield the result from false positives and false negatives.

Crucially, all these questions have to be answered without regard for what can and cannot be expected from the archaeological record. If archaeologists had access to a time machine, what kind of data would have to be collected in order to convince the rest of the scientific community that their explanation is necessarily superior to all the others? What would be the shortest scope, the maximum interval, the minimum resolution, and the minimum set of dimensions that would allow them to answer their research question and avoid the underdetermination problem? Only after having answered these questions should archaeologists ask themselves whether it is reasonable to expect the archaeological record to provide this kind of information. If the process of interest is a microscale one, and especially if it is derived from a source discipline that studies humans or animals in the present time, the answer will likely be no.

A New Program for Archaeology

If every archaeologist were to undertake the exercise above, we would witness a major recalibration of the discipline's core agenda. Archaeologists would leave the study of individual-level microscale processes and move in two opposite directions: toward fundamental cultural history and toward macroscale patterns and processes.

Figure 7.1 is a schematic illustration of the severity of underdetermination for different types of research questions. On the left side of the spectrum are low-level inferences and cultural history, which, as discussed in the previous chapter, are not heavily underdetermined by the archaeological record.

Ever since the 1960s, archaeologists have been undervaluing cultural history (Lyman, O'Brien, and Dunnell, 1997) and have shown a much greater concern for high-level explanations than for historical questions (Kintigh et al., 2014), as if, when it comes to important archaeological research, history should take the back seat (Cobb, 2014).

In the new program for archaeology, cultural history returns to the front seat. The reconstruction of cultural history is perhaps the single most important

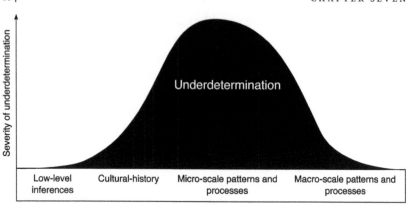

Research question

FIGURE 7.1: Schematic illustration of how, given the quality of the archaeological record, the severity of the underdetermination problem varies as a function of the type of research question.

contribution that archaeology has to offer to the social sciences. Archaeologists are making useful contributions by describing, even in the most basic manner, ancient cultures and lifeways and their distribution in time and space: What were people eating? What technologies were they using? Where did they acquire their raw materials? How fast did farming spread in Europe? When was writing invented in Mesoamerica? Some of these historical questions do involve individual-level processes, but ones that sit at the bottom of Hawkes's pyramid (fig. 6.3) and that are commensurate with the quality of the archaeological record. Over the years archaeologists have produced a wealth of knowledge about the particular history of various populations, technologies, and other traditions that is truly novel and useful to researchers outside archaeology. Just think of how our view of the human species is shaped by the knowledge that for millions of years our ancestors were hunter-gatherers and relied on chipped stone tools, and that agriculture and large-scale societies emerged independently in different regions of the world and did so only recently. And more than just useful, cultural history accounts for nearly all the epistemologically valid knowledge that archaeologists have produced. The usefulness and the validity of cultural history are the reasons why the archaeological studies that *Science, Nature,* and other high-impact interdisciplinary journals publish are ones that advance our knowledge of cultural history, especially those that push back the antiquity of cultural practices or historical events (O'Brien, Lyman, and Schiffer, 2005).

After cultural history, underdetermination increases sharply when it comes to high-level microscale processes—those that populate the theories from which archaeologists borrow their concepts and explanations (fig. 7.1 and chapter 6).

By focusing on microscale processes, archaeologists have been living beyond their epistemological means. The archaeological record, because of its quality, demands that we abandon the study of such microscale processes or at least make their study the exception instead of the norm. It demands that we cease, altogether and for good, interpreting the archaeological record in individual-level, ethnographic terms that a cultural anthropologist, a sociologist, or a field biologist would recognize. This is not because microscale processes are uninteresting or unimportant. Nor is it because they do not operate on material culture or that short-term events are invisible in the archaeological record. Rather, it is because the archaeological record is not a suitable medium for stories that unfold over individual time scales: archaeologists cannot, in most cases, isolate the action of individual microscale processes. Archaeologists must accept the limitations of the archaeological record and remain agnostic about how the vast majority of these microscale processes played out in the human past.

On the right side of figure 7.1, beyond the peak of underdetermination, lie macroscale patterns and processes. Macroscale patterns and processes are those that operate so slowly that they become visible only over time scales longer than 10^3 years and that, spatially, operate over continents, hemispheres, or the entire planet.

When you observe the night sky, the stars appear to maintain fixed positions relative to each other, unlike the sun, earth's moon, and the other planets of our solar system. But stars do move relative to each other; they just do so very slowly. So slowly, in fact, that even today we can easily recognize the same constellations that the astronomers of Bronze Age Babylonia saw thousands of years ago, such as Leo and Taurus. The motion of stars was discovered in 1718 by Edmund Halley, when he noticed that their positions in the night sky were a fraction of a degree away from where the Greek astronomer Hipparchus had placed them 1850 years earlier (Neugebauer, 1975). Thus, whereas the motion of the stars relative to each other on the celestial sphere is so slow that it is undetectable within the lifetime of any single astronomer, it is detectable with measurements with the scope of 1850 years.

What comparable phenomena, undetectable in the ethnographic present, influence human matters? We do not know, because scientists have rarely searched for macroscale patterns in human culture. Much like a geologist who wants to understand the forces that shape the earth but looks exclusively at her backyard and ignores what is beyond it, social scientists try to understand human behavior using an incredibly narrow observational window that is usually limited to orders of $<10^2$ years and $<10^2$ kilometers. The archaeological record provides us with our *only* opportunity to expand the temporal scope of our observations to capture cultural processes that, like the motion of the stars or tectonic drift, act so slowly that they are effectively invisible to social scientists, with their noses so close to the ground.

Archaeologists have yet to explore systematically human material culture at a macroscale. Indeed, the typical temporal scope of archaeological studies ranges from 10^2 to 10^3 years (fig. 6.2), and their spatial scope is usually restricted to a site, a region, or a physiographic province. What is more, archaeologists rarely work with a scope that is large *both* temporally and spatially, even when they are purportedly conducting macroscale research (e.g., the contributions in Prentiss, Kuijt, and Chatters, 2009). For instance, a long-term, "macro" study may involve the occupational history of a cave site over 8000 years—a long temporal scope indeed but confined to a very narrow spatial scope. At the temporal and spatial scope archaeologists currently work with, the effect of any macroscale phenomena would be undetectable and drowned by the noise generated by microscale processes, historical contingencies, and the forces of mixing and loss reviewed in chapters 3 and 4.

The payoffs of searching for macroscale principles in human culture could be significant. Paleontologists have made some of their most important contributions to the biological sciences by discovering patterns and processes above the species level that were not predicted by the microscale Darwinian theory. First, paleontologists leverage the vast scope of the fossil record to estimate various parameters of biological systems, such as the typical rate of evolution (e.g., Gingerich, 1983), with an accuracy that would be impossible to achieve for field biologists, who have to contend with a much narrower sampling universe. Paleontologists have also discovered several unexpected patterns and trends in biodiversity at taxonomic levels above that of the species. For example, in the early 1980s when David Raup and Jack Sepkoski first got their hands on the Compendium of Fossil Marine Families, a database of 3500 fossil marine families that lived over the last 250 million years, one of the first things they did was to plot the frequency of extinction against time. Much to their surprise, they found that extinctions were not evenly distributed over time but, rather, clumped around marked peaks (D. Sepkoski, 2012). Even more remarkable was the fact that the extinction peaks appeared to be episodic: they occurred roughly every 26 million years (Ma). After checking that their result was not an artifact of the lumping of fossil data into geological stages, of errors in taxon identification, or of the "pull of the recent," they published their results (Raup and Sepkoski, 1982, 1984) and, in doing so, started a longstanding debate that forced biologists to rethink how the history of life unfolds over geological time scales. The idea of cyclical periods of mass extinction still holds today. In 2014 Adrian Melott and Richard Bambach, using an updated version of the Compendium of Fossil Marine Families that doubled the temporal coverage of the original dataset and increased its sample size tenfold, revised the length of the cycle to 27 Ma. In addition, an even stronger signal of a 62 Ma cycle, apparently unrelated to the 27 Ma one, was also detected (Melott and Bambach, 2014; Rohde and Muller, 2005) (fig. 7.2). What drives this pattern is unknown—the smoking gun that would resolve this question has yet to be found—but the working hypotheses include vertical oscillations of our sun in the galaxy plane that modulate the flux of cosmic rays hitting our planet (Medvedev and Melott, 2007), as well as the intriguing "Nemesis hypothesis," according to which our sun has a dark, yet-unseen companion star that approaches the Oort Cloud of comets roughly every 26 Ma, pulling comets toward the earth (M. Davis, Hutt, and Muller, 1984).

Could similar cycles in global cultural diversity lie, waiting, in the archaeological record? The idea may seem far-fetched, yet analogous phenomena have already been observed in the historic records. The historian Peter Turchin mined the historical records of agrarian states and, using spectral analysis, discovered two periodic cycles of sociopolitical instability. The first one, which he calls the *secular* cycle, is two to three centuries long and is marked by waves of political instability and violence that are interspersed with periods of peace and order (fig. 7.3). This secular cycle appears in all states for which there is an accurate historical record, including those of Europe, the

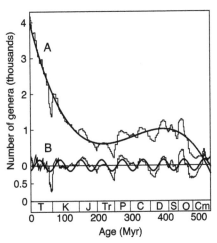

FIGURE 7.2: A 62-million-year cycle in fossil diversity. (A) The wiggly line shows the number of known marine animal genera (*n* = 36,380) but with single occurrences and poorly dated genera removed. The trend line is a third-order polynomial fitted to the data. (B) Same as A, with the third-order polynomial trend removed. The trend line is a 62-million-year sine wave superimposed on the data. (Adapted from Rohde and Muller 2005.)

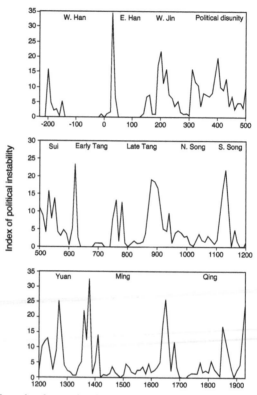

FIGURE 7.3: Secular cycles of sociopolitical instability in China. Secular cycles of sociopolitical instability with a period of 200–300 years appear in the historic record of every agrarian state and are examples of large-scale periodic cycles in human behavior. (Turchin 2012.)

Middle East, China, Southeast Asia, and the United States (Turchin, 2012). The second cycle has a shorter period of about 50 years and, though not universal, appears in many different regions, including in the United States, where outbreaks of political violence increased in frequency around 1870, 1920, and 1970 and, presumably, will do so again in 2020 (Turchin, 2012).

Paleontologists have also discovered macroscale processes, including novel evolutionary forces such as species selection, whereby species-level properties, like geographic range or population structure, affect a lineage's rate of speciation and extinction (Jablonski, 2008), as well as biogeographic drivers of global diversity that have shaped the history of life on the planet, such as tectonic-plate movements (Valentine and Moores, 1970) and continental area (Flessa, 1975).

Biogeography may also have shaped the course of human history. Jared Diamond (1997) famously argued that the shape of continents influenced the course of human societies in many important ways. He notes that ideas, crops, and technologies spread more easily between areas of the same latitude than between regions at different latitudes and with different climates and environments. Thus, innovations spread more easily within continents that extend in an east–west direction than within continents that are aligned along a north–south axis. This process, Diamond argues, explains why gunpowder spread from China to western Europe in just a few centuries, whereas the wheel developed in southern Mexico never reached the Andes. Diamond's biogeographic hypothesis is a good example of a potential macroscale driver that operates over thousands of years and at a continental level, and that is worth investigating archaeologically.

Likewise, biogeography drives linguistic diversity. There is a latitudinal gradient in the density of human languages around the world (Mace and Pagel, 1995; Nettle, 1998; Sutherland, 2003). Most of the world's languages are spoken near the equator, and language density falls as one moves away from the equator and toward the poles. This pattern is not a simple function of population density and holds true even when controlling for the area of countries. Amazingly, this latitudinal gradient in language is qualitatively similar to that found in mammal and bird species diversity such that areas with high language diversity also have high bird and mammal diversity (Mace and Pagel, 1995; Nettle, 1998; Sutherland, 2003), suggesting that the same factors underlie both patterns.

If cyclical patterns and biogeographic drivers can be detected in the historical and linguistic record, why shouldn't they also be present in the global archaeological record? For instance, does the shape and alignment of continents influence the rates of technological change? Are there biogeographic contexts that favor cultural persistence or diversity? This kind of research question falls under the umbrella of *macroarchaeology*.

MACROARCHAEOLOGY

Macroarchaeology—the search for macroscale phenomena in the archaeological record—entails a different set of research questions than the one archaeologists are trained to ask. There are very few examples of macroarchaeological studies in the extant literature, but we can draw parallels with paleobiology and the related field of macroecology (Brown, 1995; Brown and Maurer, 1989; Gaston and Blackburn, 2000) (fig. 7.4) to outline what a macroarchaeology program would look like.

Like paleobiology and macroecology, macroarchaeology can be divided into two components: (1) the search for macroscale patterns and (2) the search for macroscale processes.

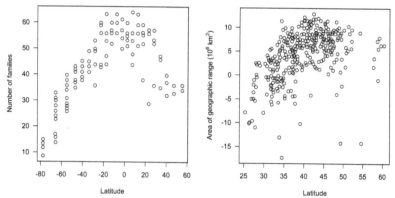

FIGURE 7.4: Macroecology is a subfield of ecology that analyzes broad statistical patterns in taxa abundance, distribution, and diversity (Brown, 1995; Brown and Maurer, 1989; Gaston and Blackburn, 2000). Though it focuses primarily on spatial patterns (as opposed to temporal ones), it illustrates the kind of research strategy that can be transposed to the archaeological record, with its focus on global databases and variables that are not species specific, such as body mass, population density, area of geographic range, and biodiversity. The graph on the left shows the effect of latitude on biodiversity. Each point represents the number of avian families for birds in the New World. (Adapted from Gaston and Blackburn, 2000.) The latitudinal richness gradient is one of the most consistent ecological patterns discovered by macroecologists (Gaston and Blackburn, 2000). The graph on the right shows the area of geographic range as a function of the latitude of the center of the range for North American land birds. (Adapted from Brown, 1995; Brown and Maurer, 1987.) The graph shows a general tendency for ranges to decrease with latitude.

Macroscale Patterns

Macroscale *patterns* are statistical signals that can be lumped into two categories: (1) temporal and spatial trends and (2) expected values.

The first category is self-explanatory. It includes temporal trends observed over long temporal and spatial scales, such as the 62 Ma cycle in fossil diversity mentioned above. Archaeologists could detect similar trends in the archaeological record. What would a plot of global cultural diversity over time look like? For example, did the number of artifact types observed in the global archaeological record increase over time? And if so, how did it increase? Linearly? Exponentially? At what rate? Did it ever reach an asymptote? Are there, superimposed on this increase, periodic cycles in global cultural diversity?

The second category of macroscale patterns, expected values, is less familiar to archaeologists. Expected values refer to the description of statistical distributions of global data in terms of central tendencies (e.g., average) and limits (minimum, maximum). The description of statistical distributions is nothing new: it is actually one of the fundamental goals of science. Indeed, much of science is not so much about testing hypotheses as about measuring properties such as the speed of light or rates of erosion and of genetic mutations.

Archaeologists have yet to seize the opportunity of aggregating archaeological contexts to measure the expected properties of various aspects of human culture. Despite more than a century of scientific archaeological fieldwork, conducted in every corner of the earth, archaeologists would still be hard-pressed to answer even the most basic questions about material culture. If a sociocultural colleague tells me that the people at her field site have been making ceramics using the same type of decoration for 200 years, I do not know if this tradition is unusually long-lived, unusually short-lived, or typical of the duration of human traditions. I may have some intuitions about it, I may be able to compare it with the ceramic traditions in the region and the time period where I work, but I cannot point her to a study that has measured the duration of cultural tradition globally and systematically. I cannot tell her, for example, if her ceramic tradition falls in the 30th or 50th or 80th percentile of traditions in terms of its duration.

Archaeologists are in a unique position to accomplish the scientific task of measuring the expected properties of human culture. As mentioned previously, archaeologists, unlike other social scientists, have access to a vast observation window. The ethnographic record represents only a sliver of human history, and even an exhaustive survey of every human society currently living on the planet would constitute but a small sample of the forms human culture can take. What is more, it is unclear to what extent human societies today, in a post–demographic transition, postindustrial, and post–Internet world, are representative of past groups. Archaeologists can cast a much wider net, sampling from a universe tens of thousands of years long, and tens of thousands of kilometers wide, allowing them to measure the average properties of cultural systems with greater accuracy than any ethnographer could ever achieve.

Some of the fundamental properties of human material culture that archaeologists can measure include the following (see also Clarke, 1968; Shott, 2015):

- The pace and direction of cultural change
  - The pace of change in material culture along linear dimensions (e.g., change in height, thickness, surface area)
  - The direction of change. Is change in one direction (e.g., toward smaller size) more likely than change in the other direction (e.g., toward larger size)? Is there a cultural equivalent to Cope's rule of size increase (i.e., the statistical trend toward larger body size over time discovered by paleontologists; Benton, 2002; Stanley, 1973)?
  - The pace of change in technological complexity. How fast do technologies increase in terms of the procedural steps they involve (Perreault et al., 2013) or in terms of the hierarchical depth and breadth of the manufacturing process (Muller, Clarkson, and Shipton, 2017)?

- The range and duration of archaeological types
  - The geographic range of archaeological types
  - The life span of archaeological types
  - The shape of the temporal frequency distribution of archaeological types. For instance, do types always rise in popularity at the same rate as when they fall out of fashion?

All these properties are statistical signals that come into view at a hierarchical level well above that of the individual. Rather than focusing on individual-level responses, like much of normal archaeology (e.g., how do individuals typically adjust their toolkit diversity in response to raw-material scarcity?), macroarchaeology is interested in the population-level properties that emerge out of the interactions between thousands of individuals over multiple generations.

What is more, the search for macroscale patterns is independent of the processes that underlie it, especially the individual-level ones. By nature, the search for macroscale patterns is descriptive. All the properties listed above can be measured while remaining agnostic about the suite of individual-level processes that explain why the values observed are what they are. Yet, by simply measuring any one of these things, archaeologists can make themselves useful and provide other disciplines with estimates for the quantities that are in their theories and models.

An Example   Let us look at the first property listed above, the pace of change of material culture. Most anthropological theories assume, without having ever tested the assumption, that cultural change is faster than biological change. The faster pace of cultural change is thought to allow humans to adapt to new environments more rapidly than other animals can, explaining why humans thrive in most of the world's terrestrial habitats. That cultural change is faster than biological change may seem self-evident and trivial, but it is not. First, remember that rates of change are inversely correlated with the time interval over which they are measured (see chapter 4). Thus, our impression that cultures change more rapidly than species could be due to the fact that we observe the objects that surround us—our cars, phones, and computers—on a daily basis, whereas when we think of biological change we tend to think of the fossil record with its intervals of millions of years. Second, the archaeological and the historical records are filled with traditions that have remained stable over hundreds and sometimes thousands of years, whereas biologists have observed, both in the laboratory and in the wild, significant phenotypic change over time periods of decades or less. At the very least, the distributions of cultural rates and biological rates of change overlap. And third, science is not about intuitions and impressions—we want to measure things, not just estimate them qualitatively.

Measuring the pace of cultural change in the ethnographic record is difficult. The pace of change of any particular tradition will depend on a host of microscale factors and contingencies that make any results difficult to generalize. What is more, rates of technological change as seen in the ethnographic present may have been affected by recent developments such as the printing press, universities, or the economic market system and as such may not be representative of rates during prehistory. It is much easier, however, to collect a sample of cultural rates of change in the archaeological record that is large enough to allow us to average over the effect of microscale factors and contingencies. This is precisely what I tried to do when I studied the rates of change in technologies in the archaeological record a few years ago (Perreault, 2012). In that study, I assembled hundreds of data points sampled from a universe that comprised the whole North American continent and the last 10,000 years. Using this large dataset, I answered two questions: (1) how fast, on average, material culture changes; and (2) how the pace of cultural change compares with the pace of biological change. Figure 7.5 shows the rates of change in material culture from the archaeological record and the rates of biological change in the fossil and historical record, both measured in darwins, d (see chapter 4).

The data suggest that the expected pace of change in the linear dimension of technologies (e.g., length, width, thickness) over a 1-year period is about 21,989 d (about 0.22%) (Perreault, 2012, fig. 2). To compare cultural rates and biological rates, I controlled for the generation time of the species in the sample, since species with shorter generation time evolve, on average, more rapidly than species with a longer life span. When controlling for the effect of generation time, I found that the pace of cultural change is faster than the pace observed in the fossil record by a factor of $e^{1.698} = 49.8$ (fig. 7.5). Thus, cultural evolution is faster than biological evolution over all observation time intervals, including time intervals that are equal to or shorter than the generation time of humans. In the biological world, species either evolve rapidly but die young or live longer but evolve slowly. What figure 7.5 demonstrates is that culture frees humans from the generation time constraints and gives us the best of both worlds: culture allows us to evolve over very short time scales that are normally accessible only to species with a very short life span, while at the same time letting us enjoy the benefits of being a species with a long life history, investing in large bodies, big brains, and long childhoods. This is what makes humans such a successful species (at least so far) and why we have come to dominate so many of the world's ecosystems.

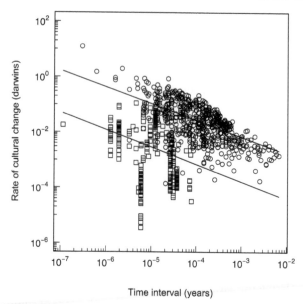

FIGURE 7.5: Rates of cultural change (circles, $n = 573$) and of biological change (squares, $n = 283$) as seen in the archaeological and fossil record. (Adapted from Perreault 2012. See the original article for details on the linear regression models fitted to the two groups of data points, but note that the scale here has been changed from a natural log scale to a log 10 scale.) The difference between the intercept of the two linear models is 1.698.

This study is only a preliminary step toward a full research program on rates of cultural change. A larger sample that includes rates from all around the world and from a wider temporal scope would not only offer us a more accurate and precise estimate of rates of change but also allow us to detect potential patterns in the pace of cultural change over time and space.

And note that this rate study does not try to explain why the rates of change of individual technologies are what they are. Instead, it seeks to average over a vast number of technological traditions and their microscale determinants in order to pick up a statistical signal, that of the typical pace of change of the material culture produced by humans. In doing so, it provides other disciplines with a useful quantity that they can incorporate in their theories and models. At the time of this writing, 95% of the publications citing the study are from fields outside archaeology, most predominately psychology, cognitive science, and biological anthropology, but also economics, genetics, animal behavior, ecology, computer sciences, physics, history, sustainability, religious studies, musicology, medical research, and, of all things, rural sociology.

Macroscale Processes

In addition to macroscale patterns, archaeologists can also identify macroscale processes that are invisible to cultural anthropologists. Macroscale processes are large-scale drivers of macroscale patterns. Like microscale patterns, they operate at a hierarchical level above that of the individual. As such, they cannot be reduced easily to microscale processes—no more than human behavior can be easily reduced to molecular interactions between the cells in our bodies. This means that unlike microscale processes, macroscale processes cannot generate variation among the individuals of the same group. Instead, macroscale processes generate variation that is detectable only over thousands of years or thousands of kilometers. Examples of research questions about macroscale drivers include the following:

- External drivers
  - Geography
    - Do the shape and the size of continents affect cultural diversity, cultural complexity, or rates of change?
    - Does latitude affect the life span of archaeological types?
    - Does latitude affect the geographic range of archaeological types?

  - Global climate
    - Do fluctuations in global climate cause changes in global cultural diversity?
    - Does climate affect cultural complexity?
    - Does climate influence the life span and geographic range of archaeological types?
    - Does the orientation of a continent's major topographic features (e.g., north–south in North America, east–west in Europe) affect the orientation of cultural geographic ranges, as it does for the range of land birds and terrestrial mammals (Brown and Mauer, 1989)?
    - Are there climatic factors (e.g., temperature, precipitation) that affect cultural traditions more than others?
    - Does the effect of climate change rate differ from the effect of absolute climate?

- Internal drivers
  - Subsistence
    - Was the advent of agriculture accompanied by changes in macroscale trends, such as a change in the typical duration of cultural traditions or in cultural diversity?
  - Technology
    - Did the advent of new materials, such as ceramics and metals, affect the life span of cultural types? Did it affect global cultural diversity?

- When a new technology arises (e.g., the bow and arrow), is the diversity
  in form concentrated in a particular part of the history of the technology
  (e.g., the beginning)? And if so, at what pace does the winnowing of less-
  efficient types occur (e.g., Lyman, VanPool, and O'Brien, 2009)?

Quasi Examples   At present, it is difficult to find an example of a study that
seeks to answer questions like the ones above. Social scientists, including archaeologists, have shown little interest in macroscale processes. And even if
they did, the global archaeological database needed to answer these questions
does not exist yet.

But there are studies that, in spirit, come close to this approach. For instance, Lyman, VanPool, and O'Brien (2009) analyzed changes in diversity
types of North American projectile points—a macroarchaeological endeavor
indeed, even though they looked at a small dataset comprising six sites from
the western United States. Another quasi example comes from the discovery
of the Neolithic demographic transition. Jean-Pierre Bocquet-Appel (2002,
2009, 2011; Bocquet-Appel and Naji, 2006) examined the demographic composition of Neolithic cemeteries and found that the spread of farming led
to an increase in fertility of populations around the world. The bioarchaeological marker of increased fertility and population growth is an increase in
a population-level property of its skeletal population: the frequency of 5- to
19-year-old individuals relative to the frequency of individuals 5 years old
and older. Obviously, the age structure observed in any particular cemetery
will be a function of many different factors, including historical contingencies, microscale processes, and forces that shape the quality of the archaeological record, including the span of time over which the cemetery was used.
To circumvent this problem, Bocquet-Appel sampled from a wide scope and
assembled a large dataset of 133 cemeteries with at least 50 skeletons. Temporally, the scope of his dataset is almost 9000 years, as the age of the oldest
cemetery in his sample is ~9000 BCE and the age of the youngest is ~350 BCE.
Spatially, the scope of his sample is equally impressive, as the cemeteries come
from the whole Northern Hemisphere—North America, Eurasia, and North
Africa (Bocquet-Appel, 2002, 2009; Bocquet-Appel and Naji, 2006). By plotting the proportion of juveniles in the cemeteries against the years since the
advent of farming in the region (fig. 7.6), Bocquet-Appel found that over the
first 1000 years following the beginning of farming in a region, the proportion
of juveniles in the skeletal population increased to 28% from 20%, on average.
By using a relative chronology (the time elapsed since the advent of farming in
a region) instead of an absolute chronology (i.e., the absolute age of the cemeteries), and by averaging over a large number of cemeteries, Bocquet-Appel

was able to pick up the faint signal of the Neolithic demographic transition. The vast dispersion of the data points in figure 7.6 is testimony to the noise generated by the hundreds of processes that operate at an individual level and influence individual fertility or that shape the quality of individual archaeological contexts. Had Bocquet-Appel worked with a smaller sample of, say, 10 cemeteries, he would have observed only statistical noise. The demographic shift discovered by Bocquet-Appel represents a major social and economic transition in the history of our species and a genuinely novel archaeological contribution to our understanding of how human populations were affected by farming.

The case of the Neolithic demographic transition is a good example of how a macroscale driver, the advent of farming, can be linked to a weak statistical pattern that is buried in noisy archaeological data and revealed by sampling from a universe that encompasses thousands of years and multiple continents. But Bocquet-Appel's work on the demographic transition differs from the macroarchaeological program outlined here owing to its bioarchaeological nature. The markers used in bioarchaeology typically have a restricted number of competing biological explanations. Here, the link between

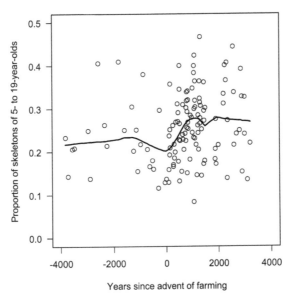

FIGURE 7.6: The Neolithic demographic transition is signaled by an increase in the proportion of skeletons of 5- to 19-year-old individuals relative to all skeletons of individuals 5 or more years old within 1000 years after the advent of farming at a location. The data ($n = 133$) come from cemeteries across the Northern Hemisphere. The line represents a locally weighted least squares regression (LOESS) fitted to the model. (Modified from Bocquet-Appel, 2011.)

the pattern observed—a change in the relative proportion of different age groups—and its putative driver, the arrival of farming, could be reduced to an individual-level process, an increase in fertility. In the case of cultural remains, however, linking a macroscale pattern to a particular microscale mechanism will always be more challenging.

APPLYING THE MACROARCHAEOLOGY APPROACH

Macroarchaeology is concerned with the characterization of statistical patterns of rates of cultural change, abundance, distribution, and diversity and with the explanations of these patterns in terms of macroscale drivers such as climate change and biogeography. This research program demands a research strategy that differs from normal archaeology in several ways:

1. *A narrow set of research questions.* Macroarchaeology, like paleobiology and macroecology, intentionally sacrifices the details and much of the information contained in the archaeological record in order for the big picture to emerge. It is about the forest, not the trees. And studying the forest means asking a narrow, but deep, set of research questions.

A single-minded focus on a set of questions that is limited and restrained by design is a defining feature of successful disciplines (Upham, 2004). In paleobiology and macroecology, such restricted sets of questions have blossomed into full-fledged, rich, and busy research programs. By pruning mercilessly its research agenda and by making the search for macroscale principles one of its main tasks, archaeology can shed its dizzying patchwork of theories and research questions and hopefully rise on the academy's ladder.

2. *A program centered on archaeological entities.* In the questions that it asks, macroarchaeology is material-culture-centric, as opposed to individual- or behavior- or social-centric. Its primary interest is in archaeological entities and their distribution in time and space. As such, it has more in common with David Clarke's *Analytical Archaeology* (1968) than it does with the contemporary and much more influential *New Perspectives in Archaeology* edited by Lewis Binford and Sally Binford (1968). Whereas the latter argues that archaeologists ought to study social, economic, ecological, and ideological processes, Clarke's approach is centered on artifacts and is concerned with the birth, growth, and death of archaeological entities. "Archaeology as archaeology" is a more appropriate motto for macroarchaeology than the "archaeology as anthropology" that has been the rallying cry of the field for six decades (Shennan, 1989).

It would be wrong, however, to see in macroarchaeology an attempt to dehumanize the past—one of the main criticisms raised against time perspectivism (Bailey, 2008). Macroarchaeology does not dehumanize the past any more than the astrophysicist who studies the shape of galaxies "de-atomizes" the universe, the field biologist who researches whale feeding behavior "de-geneticizes" the animal world, or, to use one of Bailey's examples, the pa-leobotanist reconstructing Pliocene vegetational history "de-botanizes" the study of ancient plant life by failing to demonstrate that Pliocene plants used photosynthesis for energy (Bailey, 2008, 23). It would be surprising if prehis-toric people did not live rich lives in which social norms, culture, perception, identity, power, and agency played crucial roles. But not everything that may have conceivably taken place in the human past can, and ought to, be re-constructed (Bailey, 2008). Like time perspectivism, macroarchaeology does not move away from an individual-centered research program for dogmatic reasons; rather, it does so out of epistemological necessity.

In the preceding sections, I have focused on one of the most basic units in archaeology, the artifact type. The artifact types that macroarchaeology is pri-marily interested in are the sets of homogeneous populations of artifacts "that share a consistently recurrent range of attribute states" (Clarke, 1968, 206) and that have a unique spatiotemporal range and represent heritable continuity and cultural traditions (Lyman, VanPool, and O'Brien, 2009; O'Brien and Ly-man, 1999). Such units have already been constructed by archaeologists to measure time, and they can be repurposed for the task of macroarchaeology (Lyman, VanPool, and O'Brien, 2009; O'Brien and Lyman, 1999). I set aside here the methodological issues surrounding the creation of archaeological types because they have been discussed at length elsewhere (e.g., Clarke, 1968; Dunnell, 1971; Lyman, VanPool, and O'Brien, 2009; O'Brien and Lyman, 1999, 2002, 2003), but it is worth emphasizing that macroarchaeology does not re-quire an essentialist view of types, no more than the analysis of fossil diversity shown in figure 7.2 implies that species are essential objects.

Macroarchaeology, however, can encompass analytical units constructed at a variety of other hierarchical levels. Although the effect of mixing and loss on correlations in archaeological contexts would complicate such an analysis, macroarchaeology could be applied to the study of archaeological culture—that is, those sets of types that consistently appear together in as-semblages within a limited geographic area (Clarke, 1968, 247). For example, archaeologist Katie Manning and her colleagues (Manning et al., 2014) have collected chronometric data about the archaeological cultures of Neolithic Europe. While their primary goal was to refine the chronology of these cul-tures, they also discovered an interesting macroscale pattern in the shape of the temporal frequency distribution of radiocarbon dates of the cultures: they are normally distributed, much like the waxing and waning popularity of ar-chaeological types or even of marine invertebrate genera (Foote, 2007). Some follow-up questions to this finding include whether or not the archaeological cultures in other parts of the world also rise and fall following a Gaussian pattern, but also the same set of macroarchaeological questions that can be asked about archaeological types: questions about their duration, their geographic range, as well as about the external and internal drivers of global diversity in archaeological cultures.

3. *General properties.* Macroarchaeology analyzes general properties that can be measured, at least theoretically, at any given point in time and space in the human past. Archaeologists are used to building datasets with time-specific, place-specific, or technology-specific variables: lithic data, zooarchaeological data, ceramic data, household architecture data, and so on. In contrast, macroarchaeology is about drawing inferences from the statistical distributions of variables among many different traditions and technologies from different times and places. The difference between normal archaeology and macroarchaeology is analogous to the difference between a zoologist studying bat echolocation systems (a trait that is species-specific since not every species has a capacity for echolocation) and a macroecologist analyzing the geographic range of terrestrial species (a trait that is not species-specific since every species has a geographic range). Some of the general analytical variables of interest to macroarchaeology include temporal ranges, geographic ranges, diversity, complexity, rates of change, rates of appearance, and rates of disappearance.

These variables have the advantage of being observable directly in the archaeological record. In contrast, the variables that populate normal archaeology, even those that are general and universal, are based on indirect proxies in material culture. For instance, a comparative study of complex societies may look at the relationship between variables such as population size, number of administrative levels, social network topology, or wealth redistribution mechanisms—all measurements that are based on unverified and unverifiable inferences.

4. *Large databases with wide spatial and temporal scope.* The "macro" in macroarchaeology refers, first and foremost, to its scope (as opposed to the hierarchical level of the analytical units, as it is sometimes used; see, e.g., Prentiss Kuijt, and Chatters, 2009). Macroarchaeology takes a 10,000-miles view of the archaeological record. This translates into datasets that have a much larger scope than archaeologists typically analyze. In fact, by embracing macroarchaeology, archaeologists would be doing the very opposite of what they have been trying to do for years: moving as far away as possible from an ethnographic scale of analysis.

The secret to paleontologists' success was the "crunching of the fossils" (D. Sepkoski, 2012; Turner, 2009), that is, the analysis of global multitaxa databases. Starting in the 1970s, Jack Sepkoski started to assemble the Compendium of Fossil Marine Families (J. Sepkoski, 1982), the first global and comprehensive database of fossil marine animals. The database was simple—it contained the times of origination and extinction of the different families of marine animals. Yet, the database was enough to identify several temporal trends in biodiversity and extinction rates. Today, the database exists under the name of Paleobiology Database (www.paleobiodb.org) and contains hundreds of thousands of data points. It is this database that has allowed paleontologists to replace the interpretation of the history of individual taxonomic groups in microevolutionary terms by a true search for macroscale patterns in biodiversity. Archaeologists too can build global archaeological databases—large databases that pool together hundreds of analytical units drawn from a vast sampling universe, with a scope that is on the order of at least 10^3 years and 10^3 kilometers or, ideally, that encompasses the global archaeological record, both spatially and temporally.

Macroarchaeology and Underdetermination

Macroscale patterns and processes are less likely to be underdetermined by the archaeological record than microscale ones. The macroscale patterns observed are less likely to be false ones, and the macroscale processes identified are more likely to be the right ones.

First, the number of hypotheses that compete at a macroscale is smaller than at a microscale. As the temporal and spatial scale at which a pattern emerges increases, the number of processes that can explain the pattern decreases. For instance, there are a myriad of possible explanations for the function of a plaza in an ancient city or for why two prehistoric houses differ in size. But there are few explanations for, say, millennial-scale fluctuations in global cultural diversity over the last 500,000 years. Similarly, in Bocquet-Appel's study of the Neolithic demographic transition described above, a difference in the age ratio between two cemeteries may be due to a plethora of factors (the temporal resolution of the cemeteries, migration, changes in social norms, short-term fluctuation in climate, or sample size). But a similar pattern observed across more than one hundred cemeteries spread across the entire Northern Hemisphere can be accounted for by very few factors besides something like the advent of farming. Indeed, any explanation has to be commensurate with the time and spatial range of the pattern observed. Just as the emergence of complex societies over the last five thousand years happened too late in human history to have been solely the result of individual-level psychological processes or population growth (Richerson and Boyd, 2005), most processes known to the social sciences operate too rapidly to viably explain macroscale archaeological patterns.

The second reason why macroarchaeology reduces the underdetermination problem is that global archaeological databases would have a quality that is commensurate with macroscale patterns and processes. A global archaeological database would have the scope necessary to make macroscale phenomena visible (chapter 2). And just as important, it could act as a low-pass filter that cancels the noise generated by microscale factors. The noise-canceling property of global databases hinges on the virtues of low-resolution data. When studying microscale processes, the time averaging and space averaging of archaeological assemblages are a nuisance. But when it comes to studying macroscale processes, they are a virtue since they iron out the noise created by microscale processes and reveal the large-scale trends (Bailey, 1981, 2008; Barton and Riel-Salvatore, 2014; Fürsich and Aberhan, 1990; Higgs, 1968; Lyman, 2003; Olszewski, 1999; Stern, 1994; Wilson, 1988).

There is nothing esoteric about the noise-canceling property of global databases. It depends on a simple statistical principle that we are all familiar with. For instance, when an instructor calculates the average test score in her class, she is effectively trying to mute the various individual-level factors that operate over time scales that are shorter than a semester and that can influence how well a given student does on an exam—how motivated a student is about the subject topic, how much time he spent studying, the studying technique he used, his age, or whether or not he partied the night before. By muting these factors, the instructor is trying to unravel the signal of interest: how much a cohort of students has learned over the semester. Similarly, when we fit a linear regression to a cloud of points, we are using the noise-canceling properties of aggregate data. A linear regression model assumes that the signal of interest, the effect of the predictor x on variable y, can be represented as $y = x + \varepsilon$, where ε is the noise generated by all the other factors that are independent of x but that also affect y. As the scope of a dataset and its sample size increase, the effect of ε on the estimation of the mean shrinks; that is, the signal-to-noise ratio improves.

By canceling the noise that microscale processes generate, a global database would effectively allow archaeologists to control for them. A global database would be so large in terms of sample size that the microscale processes that operate within its scope cancel each other out—or at least cancel each other out to a sufficient extent for the signal of macroscale principles to be detected. The key here is that the different microscale processes affecting material culture do so in different, and sometimes opposite, ways: a process may act, for instance, to increase the number of ceramic styles in a group while others will act to decrease it. When a sample size is large enough, the microscale processes pulling in opposite directions end up canceling each other out. This phenomenon explains why bell-shaped curves are common in nature. For example, the distribution of stature is bell shaped, even though stature is influenced by a great many factors, none of which are random. And yet, stature converges to a normal distribution because fluctuations away from the mean in one direction are, on average, canceled out by fluctuations in the other direction. Comparing the distribution of stature between two countries is, effectively, a way to control for the multitude of within-country processes that affect stature, in order to reveal the effect of macro-, country-level properties, such as GDP or health care system. A global archaeological database would work the same way. Archaeologists may not be able to single out individual microscale processes or to control for them individually, but they can control for them in bulk, in the aggregate of global databases.

Global archaeological databases would improve the signal-to-noise ratio in different ways. Subtle patterns that are lost in the background noise when observed at a scale of 10^0–10^3 years and 10^0–10^3 kilometers can become visible when the scope of observation is increased to a macroscale. For instance, a linear trend may be too small relative to the effect of microscale processes to be visible in a dataset of 200 sites sampled from a 2000-year period (a large dataset by current archaeological standards), but become visible even to the naked eye in a global database of 2000 sites sampled from the last 20,000 years (fig. 7.7; see also fig. 3.15A). Another powerful way to mute the effect of microscale processes and reveal a macroscale pattern is to use signal averaging. In signal averaging, a set of repeated measurements is averaged in order to increase the strength of the signal relative to the noise. Signal averaging is primarily used to study radio signals, but it can be applied to archaeological contexts. For example, a time series of archaeological measurements (e.g., diversity) taken from sites in one region shows no discernible pattern and looks like random noise (fig. 7.8, left panel). What is more, the time series from different regions of the world appears to show no correlation between the measurements (fig. 7.8, middle panel). And yet, hidden in all that noise, lies a macroscale pattern. Averaging the measurements across all the regions at different points in time filters out this noise, revealing periodic cycles in the time series (fig. 7.8, right panel).

By building global databases, archaeologists can approximate the conditions of randomized controlled trial experiments—the gold standard of human

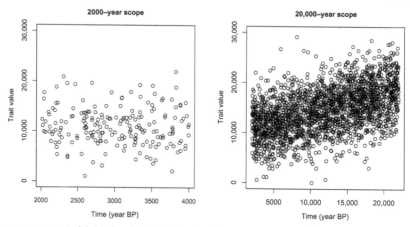

FIGURE 7.7: A global archaeological database has the scope necessary to make macropatterns visible. A linear trend ($y = x0.2$) is invisible in a dataset with a scope of 2000 years (left, $n = 200$) both because the trait value decreases slowly with time and because of the noise generated by microscale processes, modeled here as random noise normally distributed around a mean of 0 and standard deviation of 4000. The trend, however, is manifest in a global database with a 20,000-year scope (right, $n = 2000$).

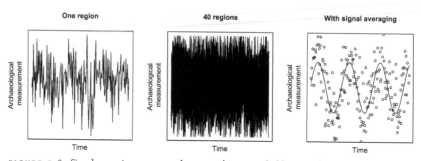

FIGURE 7.8: Signal averaging can unravel macroscale patterns hidden in archaeological data. Left: a time series of archaeological measurements taken from the archaeological sites in a region. The archaeological measurements follow a periodic cycle ($y = \sin x$), but this cycle is buried by the noise generated by microscale processes (random numbers drawn from a normal distribution with mean 0). Middle: because of the noise in the data, the time series of measurements from 40 different regions appears to show no correlation between the measurements, and no patterns are visible. Right: Signal averaging filters out the noise and reveals the periodic cycles in the time series.

research. Randomized controlled trial experiments (chapter 2) are a powerful way to deal with covariates and reduce the chance of obtaining a false result. For instance, participants in a clinical trial may be assigned, randomly, to one of two conditions: a control group, which receives a placebo, and an experimental group, which receives the new drug. When the size of the two groups is large enough, the two groups are similar in every aspect (age, genes, diet, life history, etc.) but one: the presence of the drug. Because there are no other systematic differences between the two groups, any statistical difference in health outcome between the two groups can be attributed to the effect of the drug and only to it. Similarly, global databases can be used as "treatment" conditions when they are so large that they differ in no systematic way but for the treatment condition. For example, a difference in global cultural diversity between periods of cold and warm global climates found by measuring diversity in thousands of sites sampled from the entire archaeological record can be attributed only to a difference in climate. The same method could also be used to study the effect of the shape of continents or of latitudes on different aspects of the archaeological record.

For the same reasons that it controls for microscale processes, macroarchaeology alleviates the issue of false patterns generated by the forces of mixing and loss discussed in chapters 3 and 4. When archaeologists sample, as they often do, from universes that are less than a thousand years long and have a spatial scale ranging from a single site to a physiographic province, the effects of mixing and loss can be strong relative to the anthropological signals contained in the record, and there is a good chance that the patterns they see are false ones. With the macroarchaeology approach, however, sample sizes are so large, and the sampling universes so vast, that many of these false patterns will disappear.

Of course, the use of large samples to improve statistical signals is not new. Archaeologists have been dealing with issues of sample size and sampling errors for decades. The novelty of the macroarchaeology approach lies in using databases that have temporal and spatial scopes that are several orders of magnitude larger than the typical datasets currently used by archaeologists.

The scope and the sample size necessary to unravel macroscale patterns and processes need to be very large, both temporally and spatially, because the array of factors that influence human behavior is very large—larger in fact than that of any other animal species. This is evidenced by the fact that the behavioral variation among humans, both within and between societies, is unmatched in the animal world. This unmatched behavioral variation is due to two things. First, our species inhabits a uniquely large array of ecological and social environments. And second, human behavior is deeply influenced by cultural information acquired through social learning. This reliance on culture means that human behavior is path dependent—how an individual behaves depends in part on how the individuals from her parent's generation behaved—which sets different societies on different historical paths. Indeed, the effect of culture is so potent that cultural history is a better predictor of the behavioral variation between human societies than ecological habitat. With my Arizona State University colleague Sarah Mathew, I compared the relative effect of environment and cultural history in explaining the behavioral variation among 172 tribes of western North America at the time of European contact (Mathew and Perreault, 2015). These tribes occupied a vast array of ecological habitats, ranging from desert to tundra, and also had different cultural histories, as evidenced by the fact that they spoke 116 distinct languages. In our analysis, we compared the extent to which the behavioral variation among the tribes is explained by variation in ecological environment and by cultural history. We found that cultural history is a better predictor of whether or not a tribe possesses a cultural trait than ecological habitat for a wide range of cultural traits such as technology and material culture, marriage and family organization, economic organization, ceremonies and rituals, supernatural beliefs, kinship system, political organization, warfare, settlement patterns, and sodalities (Mathew and Perreault, 2015). We also detected the effect of cultural ancestry over thousands of years. These results mean that despite what archaeologists frequently assume, ecological habitat alone is not a good predictor of a group's economy, social organization, or even subsistence patterns. Two groups may live in the same environment but, if they don't share the same cultural ancestors, may behave in different ways. Similarly, two groups that are culturally related may behave similarly even though they live in different habitats.

Culture affects every aspect of human behavior, including those that were assumed to be universal features of human psychology. Much of the psychological and behavioral science research conducted during this and the last century used subjects from Western industrialized countries, most of whom were US undergraduate students. What is more, researchers often assume that their findings are universal: findings from undergraduate students are extended to the whole species. Psychologist Joseph Henrich and his colleagues (Henrich, Heine, and Norenzayan, 2010) recently evaluated this assumption using a cross-cultural database of experimental results. What they found was surprising, to say the least: human societies vary considerably even along domains that, intuitively, we expect to vary little cross-culturally: economic decision making, spatial cognition, and even visual perception (the textbook-classic Müller-Lyer illusion). Not only that, but Henrich et al. found that WEIRD societies (Western, educated, industrialized, rich, and democratic) are more than just unrepresentative of the human species: they are significant outliers. Psychologists have some homework to redo.

The study by Henrich et al. should give archaeologists a pause. Along with the study on western North American tribes, it amplifies the critique of ethnoarchaeology that generalizing from ethnographic societies to the archaeological record is a risky proposal. The two studies also speak to the importance for archaeologists of sampling widely and largely. To control for the effect of cultural historical trajectories, archaeologists need to cast a net wide enough that it will encompass multiple historical lineages. Many, if not all, macroscale processes will emerge only at a scale well above the level of society or ethnolinguistic group.

This is one of the reasons why processual archaeologists failed to identify meaningful ahistorical principles of human behavior. Their scope of observation, however large they thought it was, remained too narrow to iron out the effect of microscale factors, historical contingencies, and the forces of mixing and loss. Even today, the typical sampling universe that archaeologists draw from (fig. 6.2) is not large enough to mask the noise generated by microscale drivers and to reveal macroscale principles, no more than the average score of a handful of students taken at random from a classroom of a 100 will be representative of the true class average.

All that is not to say that macroarchaeology is immune to false results and underdetermination. There are indeed systematic biases and processes that can give rise to macroscale patterns. For instance, age-biased preservation loss will generate a pull of the recent whereby global cultural diversity is expected to be skewed toward the present.

Archaeologists can alleviate these issues in different ways. To identify the biases skewing their data they can compare data from well-preserved and poorly preserved archaeological traditions or compare datasets that represent different analytical and classification standards (e.g., from different parts of the world or assembled at different points in time) (Foote, 1996). Once these biases and the magnitude of their effect are known, they can be included in the statistical models used to detect macroscale phenomena. Similarly, taphonomic biases that are known to exist can be accounted for in the sampling procedure used to build datasets (e.g., Olszewski, 1999). Or new statistical methods can be developed. For instance, paleontologists place confidence intervals on the stratigraphic range of fossil horizons in order to solve the problem of systematic underestimation of time ranges. The methods used to do so could profitably be adapted by archaeologists. These methods assume that the gap between the known end of a range and its true end is just a gap like any others and also results from preservation and observational losses. This means that the distribution of gaps within the range carries information about the two gaps at the beginning and the end of the range. The methods can thus be used to place confidence intervals on ranges (Marshall, 1990, 1994, 1997; Strauss and Sadler, 1989). This implies that the more fossil remains from a taxon that are known, the greater is their inferred preservation potential and the more confident we can be that the known range of the fossil remains is close to their true range—that is, a lack of preservation in older or younger contexts is more likely to represent a true absence (Foote, 1996).

Archaeologists can also remove analytically the effects of systematic biases. This last solution requires archaeologists to ask not merely what macropatterns exist in the archaeological record but what macropatterns exist *above and beyond* the patterns expected to arise from, say, the pull of the recent. For instance, in figure 7.2, the researchers detected periodic cycles in biodiversity after removing the effect of poor data (curve A) and after removing the pull-of-the-recent bias (curve A to curve B). Similarly, in my study of archaeological rates of change (Perreault, 2012), I factored out the fact that the temporal sampling interval of the archaeological record increases the farther we go back in time by plotting the rates of change against the time interval over which they are measured.

CONCLUSION

Archaeologists know virtually nothing about human culture and behavior at a macroscale. The macroarchaeology approach, with its vast scope, examines humans from a vantage point that is so removed from the way we experience the world in our daily lives that it is difficult to even imagine what macroscale patterns and processes should look like. What is more, archaeologists cannot rely on other social disciplines to provide them with hypotheses to test. At least initially, the search for macroscale principles in the archaeological record will be exploratory, empirical, and largely unguided by theory and predictions.

Most of the macroscale processes identified by archaeologists are likely to be external (Bailey, 1983, 2007, 2008). When paleontologists recalibrated their research program to the quality of the fossil record (chapter 6), they abandoned for the most part interpretation of the fossil record in terms of biotic forces (i.e., internal forces) and replaced it with a search for abiotic drivers (i.e., external forces). These abiotic drivers included climate change, oceanography, tectonic events, asteroid impacts, changes in ocean circulation, and various other aspects of the physical-chemical environment (Jackson and Erwin, 2006)—processes that leave clear signatures in the geological record and that are independent of the fossil record. Paleontologists did not switch from an internal perspective to an external one because they viewed biotic forces as unimportant. Rather, they simply recognized that internal forces, such as species competition, do not leave unambiguous physical traces and are underdetermined by the fossil record.

Macroarchaeologists will probably have to embrace externalism too. Not because external forces are more important determinants of human behavior, but rather because they may be the only class of forces that can be studied archaeologically without being underdetermined (Bailey, 1983, 2007, 2008).

First, external drivers such as climate, being physical phenomena by nature (as opposed to social or psychological), are more likely to leave unambiguous traces in the field that can be observed and measured directly and without recourse to unverified, indirect proxies.

Second, external drivers can operate over very long time scales and spatial areas. Processes that are internal to the human species, such as psychological or social factors, operate at the hierarchical level of the individual and, by nature, within the span of a human lifetime—much too rapidly to give rise to macroscale patterns. In contrast, external forces are not bounded by the human life span and are free to operate over large scales that are commensurate with the scope, the sampling interval, and the resolution of the global archaeological record.

Third, external drivers are independent of the archaeological record. Internalist arguments in archaeology always run the risk of being circular, because there is no way to demonstrate that the internal driver inferred by an archaeologist is truly independent from the dependent variable. For instance, was the militaristic iconography observed on the ceramics of a state society used by the ruling elite to legitimize their military expansion, or were both iconography and military expansion simply covarying with a third, unknown phenomenon? Was the display of wealth in burials used as a political strategy, or was the causal arrow pointing in the other direction, with burial treatment reflecting the wealth and political capital individuals acquired over their lifetimes using other political tools? Or was it both? In comparison, the external approach largely dodges this circularity problem. The shape of continents may have affected human history, for example, but human history did not affect continental shape (i.e., at least until the sea level rises as a result of human-activity-driven climate change).

Unless archaeologists put the fruits of their labor into a single integrative global archaeological database, much of the discussion about macroscale patterns and processes—what they could be, or even whether they exist or not—will remain speculative. But remember that the search for macroscale patterns and processes is only one component of the new agenda for archaeology. The other component, the reconstruction of cultural history, does not hinge on the existence of macroscale processes to be a valid pursuit, nor does the need to abandon the study of microscale processes that are underdetermined by the archaeological record.

8

Final Words

The goal of many archaeologists is to interpret the archaeological record in terms of microscale processes. In embracing this goal, archaeologists have uncritically borrowed a programmatic agenda that was designed by, and for, researchers who study humans in the present time and who use data that have a scope, a sampling interval, a resolution, and a dimensionality that are orders of magnitude different from what archaeologists have access to.

In doing so, archaeologists have been, unwittingly, publishing results that have to be, for the most part, wrong. They are offering explanations for the human past that are merely consistent with the record instead of being supported beyond a reasonable doubt by a smoking gun. It is no wonder that archaeology plays, as Geoff Bailey puts it in the quotation that opens this book, a minor role among the disciplines that study humans: archaeology suffers from an inordinate underdetermination problem.

This underdetermination problem stems in part from the fact that archaeologists lack a theory that describes, mechanistically, the various pathways that lead to underdetermination (chapter 1). They also lack a theory that links these pathways to measurable aspects of the quality of the archaeological record (chapter 2) and that articulates how these different aspects of the quality of the archaeological record are shaped by various forces, such as cultural deposition, sedimentation, or fieldwork technique (chapters 3 and 4). In addition, archaeologists have put little effort so far in measuring empirically the quality of the archaeological record (chapter 5). And the history of the discipline, the way archaeologists understand the principle of uniformitarianism, the anthropic bias in their view of human behavior, and the way they are trained to confirm hypotheses have allowed archaeologists to shield themselves from having to recognize the underdetermination problem that plagues their research (chapter 6).

One important logical conclusion emerges from these six chapters: most microscale processes are irremediably underdetermined by the archaeological record. Those that are not are the exceptions, not the rule. Of course, nothing stops archaeologists from continuing to couch their interpretations in microscale terms. But nothing can change the fact that we will *never* know, beyond anything close to a reasonable doubt, whether these interpretations are right or not. We will never know beyond a reasonable doubt what caused the appearance of complex behaviors during the Middle Stone Age in Africa, what the meaning of the Chauvet Cave paintings was, what mobility strategy the first foragers to occupy the Tibetan Plateau utilized, the effect of prehistoric taboo on faunal assemblages, what social function the large feasts held on the Pacific Northwest Coast served, whether or not the foragers of the Great Basin optimally adjusted their diet to their environment, what role ancestry lineages played in the social stratification at the site of Çatalhöyük, how perceptions of personhood evolved during the pre-Classic period in the American Southwest, whether funeral gatherings in Siberia were used as opportunities to garner political support, what the nature of the social dynamics at the frontier of southern Peru during the Early Intermediate period was, the degree to which the infrastructures of the ancient cities of India represent a consensus between their builders and their users, whether territorial expansion in early state society was associated with the delegation of authority to local administrators, whether cultural group selection played a role in the rise of state societies in China, or even why state societies emerged in the first place. We will never know any of these things for the exact same reasons that we will never know what song Ötzi the Iceman liked to sing or what name he responded to. The archaeological record is, quite simply, an inadequate source of information to research any of these topics. It will always be unproductive to generate and dwell on hypotheses that will remain, forever, just that—hypotheses. If archaeology is to be a science, it needs to stop asking unanswerable questions, no matter how interesting they may be.

Yet, the study of microscale processes is so deeply ingrained in the practice of archaeology that it is hard to imagine what is left if not for them. What is left is a two-pronged research agenda: archaeologists can continue to engage in the reconstruction of cultural histories and they can start to search for macroscale patterns and processes in the global archaeological record.

Cultural history may not involve the kind of high-level explanations that archaeologists are so fond of, but it remains a complex, challenging task that goes well beyond the mere tallying of artifact types and dates (Tolstoy, 2008). And the good news is that archaeologists have been reconstructing cultural history for more than a century and have developed sophisticated theories and methods that have produced some of the most exquisite data on the history of our species.

The study of macroscale patterns and processes, on the other hand, is uncharted territory. The archaeologist Geoff Bailey, pondering on the difficulty of adopting a long-term perspective on human behavior, remarked that to do so requires us to enter an alien intellectual landscape where the familiar landmarks and signposts are missing (Bailey, 2008, 29). And there are no compasses to guide us—the current theories in the social sciences have very little to say about what long-term trends could exist in the global record of human material culture or what macroscale drivers may have shaped the course of human history. Archaeology needs pioneers to explore this brave new world, work out the map, and report on the novel and possibly theory-challenging discoveries they make.

Searching for macroscale principles will require archaeologists to change the way they do things. A large global database of cultural traditions, similar to paleontology's Paleobiology Database, will need to be assembled before patterns and trends in cultural diversity can be detected in the archaeological record. Much of the information that would populate such a database already exists, but it is dispersed in academic books, journals, and field reports. Until all this information is integrated in a global database, archaeologists will keep underutilizing the data they spend so much time and effort collecting in the field. How many macroscale patterns and processes are waiting to be discovered, not in the field, but in the massive trove of data archaeologists have produced over the years?

For years, archaeologists have used the archaeological record as if it were a window on the past—as if they could look through it, like an observer behind a one-way mirror, and study past human societies the way cultural anthropologists do present ones. But the archaeological record is no window on the past. There are no ethnographic pictures for us to see in it, no more than on a Rorschach inkblot card. A more appropriate metaphor for the archaeological record is that of echoes from the past (T. Murray and Walker, 1988, 277). It is not a wiretap—no whispers or conversations can be heard from it. Rather, it is a cacophony of distorted and mixed sounds that have been reverberating for millennia and that reach us well after the original sounds have stopped. But it is nonetheless a signal of human activity, one that can be transformed and made meaningful. It is up to us to figure out what we can do and, just as important, what we cannot do, with that signal. Changing archaeology's research agenda will not be easy—old habits die hard. But the rewards to be reaped are huge: archaeologists can make contributions that are truly valid, novel, and useful, and they can take a seat at the high table of social sciences. Only by recognizing the quality of the archaeological record can we transcend its limitations.

Appendix A

A Formal Model of the Effect of Mixing on Variance

Imagine a continuous archaeological trait, X. As in figures 3.12A and B, the mean of trait X changes over time, but its variance remains stable. Thus, the value of any single measurement of that trait can be expressed as the mean of the context from which it is drawn, plus some deviation from the mean (Hunt, 2004). Similarly, in a mixed assemblage, the value of an observation is the sum of two variables, M and D, where M is drawn from the population of means and D is drawn from the population of deviations from the group mean (Hunt, 2004):

(A1) $X = M + D.$

When M and D are uncorrelated, as they are in the case of the normal distribution, the variance of the trait, V_x, is the sum of the variance among the means, V_M, and the variance of the trait, V_D (Hunt, 2004):

(A2) $V_x = V_M + V_D.$

When the distribution of the archaeological trait is temporally autocorrelated, such a system can be modeled as a Markovian process (Hunt, 2004), in which the mean of the trait at any given point in time will be the mean at the previous time point, plus or minus an increment of change:

(A3) $M_i = M_{i-1} + \gamma.$

where M_i is the mean trait value at time i, M_{i-1} is the mean trait value at the previous time step, and γ is the amount of change in mean value that took place over the two time steps.

In turn, γ can be modeled as a variable drawn randomly from a distribution with a mean μ_{step} and a variance δ_{step}.

Variance inflation due to mixing thus depends on the particular combination of μ_{step} and a variance δ_{step}. The expected inflation of the variance due to mixing is thus

(A4)
$$E[V_M] = \frac{t^2 \mu_{step}^2}{12} + \frac{t\delta_{step}^2}{6}$$

(see Hunt, 2004, for full derivation).

Appendix B

Source of Time Intervals and Time Resolutions from Journal Articles

The dataset is available upon request from the author.

Journal and source	*n* time intervals	*n* time resolutions
Current Anthropology		
Golovanova et al., 2010	4	
Bettinger et al., 2010	5	6
Sanhueza and Falabella, 2010	12	
Kansa et al., 2009	3	3
Miller, Zeder, and Arter, 2009	5	6
A. Smith and Munro, 2009	6	7
B. C. Finucane, 2009	11	
N. Roberts and Rosen, 2009	4	
Vega, 2009	8	
Neme and Gil, 2009	6	7
Ladefoged and Graves, 2008	39	
Assefa, Lam, and Mienis, 2008	1	
Flad, 2008	14	15
Harrower, 2008	7	
Szabó, Brumm, and Bellwood, 2007	6	
D. Jackson et al., 2007	4	
Frink, 2007	3	
Bentley et al., 2007	7	7
Nami, 2007	8	
Kohler and Turner, 2006	5	6
Monnier, 2006	68	
Haas and Creamer, 2006	24	
Borrero and Barberena, 2006	39	
Sealy, 2006	66	
Kolb, 2006	68	

(*continued*)

(continued)

Journal and source	n time intervals	n time resolutions
Robin, 2006	2	2
Coltrain, Hayes, and O'Rourke, 2006	76	
Garcia Guix, Richards, and Subir, 2006	10	
Bocquet-Appel and Naji, 2006	46	
Adler et al., 2006	9	2
James and Petraglia, 2005	11	
Russell, Martin, and Buitenhuis, 2005	2	3
Bandy, 2005	7	10
Sutter and Cortez, 2005	4	
Robbins et al., 2005	7	
Rosen et al., 2005	10	
McNabb, Binyon, and Hazelwood, 2004	2	
Gibaja Bao, 2004	1	
Friesen, 2004	14	
Munro, 2004	3	
Pinhasi and Pluciennik, 2004	11	
de la Torre, 2004	1	
B. Adams and Ringer, 2004	5	
Shimada et al., 2004	9	
de Beaune, 2004	8	
Lucero, 2003	42	
Arrizabalaga et al., 2003	3	
Richards, Price, and Koch, 2003	14	
Gil, 2003	8	
Dizon et al., 2002	3	
Tykot and Staller, 2002	3	
Barham, 2002	5	
Brantingham et al., 2001	5	
Sealy and Pfeiffer, 2000	76	
Benz and Long, 2000	3	
Stiner, Munro, and Surovell, 2000	17	
Varela and Cocilovo, 2000	3	4

Journal of Archaeological Research

Ur, 2010	13	14
Yao, 2010	4	5
Blitz, 2010	7	
Beekman, 2010	5	
Zeder, 2009	65	8
Tartaron, 2008	27	29
Kuzmin, 2008	17	
Parkinson and Duffy, 2007	11	1
Zilhão, 2007	64	
Vaughn, 2006	8	9
Pool, 2006	9	10
M. Smith, 2006	2	3
Balkansky, 2006	9	10
Byrd, 2005	147	

(*continued*)

Journal and source	*n* time intervals	*n* time resolutions
Cooke, 2005	41	4
Hoopes, 2005	26	
Wells, 2005	10	
Schroeder, 2004	27	
Sassaman, 2004	31	6
Janusek, 2004	10	7
Rothman, 2004	7	8
King, 2003	6	7
Robin, 2003	2	3
Hegmon, 2002	8	9
Lambert, 2002	43	
Stiner, 2002	26	
Erlandson, 2001	77	
Savage, 2001	7	9
Wilkinson, 2000	34	36
Keegan, 2000	3	3
Fiedel, 2000	53	
World Archaeology		
Frachetti et al., 2010	4	5
Shapland, 2010	3	4
MacKinnon, 2010	2	3
Barber, 2010	5	
McCoy and Graves, 2010	5	
Zangrando, 2009	10	
V. Thompson, 2009	1	2
Angelucci et al., 2009	10	
Bailey and Galanidou, 2009	26	
Arias, 2009	4	
Sulas, Madella, and French, 2009	2	3
Braemer et al., 2009	11	
Fuller and Qin, 2009	22	7
Teyssandier, 2008	3	
B. Roberts, 2008	12	
Iriarte, 2006	12	
Brantingham and Xing, 2006	12	
Efstratiou et al., 2006	3	
Walsh, Richer, and de Beaulieu, 2006	9	
Janusek, 2006	2	3
R. Adams, 2006	3	2
Garcea, 2006	4	3
Pearson, 2006	18	6
Matsui and Kanehara, 2006	1	
Walde, 2006	2	3
Gibson, 2006	11	
Rosenswig, 2006	5	6
Fairbairn, 2005a	6	2

(*continued*)

(continued)

Journal and source	*n* time intervals	*n* time resolutions
Bogaard, 2005	5	6
Fairbairn, 2005b	1	
Denham, 2005	2	3
Fiedel, 2005	15	
Chatters and Prentiss, 2005	9	6
J. Arnold and Bernard, 2005	17	3
Varien and Ortman, 2005	12	
Neves et al., 2004	4	
Lahiri and Bacus, 2004	17	17
Ray, 2004	19	
Singh, 2004	11	13
Srinivasan, 2004	11	4
E. Adams, 2004	1	2
Woodward and Woodward, 2004	5	5
Blom and Janusek, 2004	2	2
Bell and Renouf, 2003	1	2
Van de Noort, 2003	6	
Barber, 2003	1	2
O'Sullivan, 2003	45	
Keita, 2003	2	3
Warrick, 2003	8	
Steyn, 2003	3	3
Bright, Ugan, and Hunsaker, 2002	15	
Schepartz, Miller-Antonio, and Bakken, 2000	13	
Shoocongdej, 2000	6	
Latinis, 2000	23	
Tayles, Domett, and Nelsen, 2000	2	3
Bulbeck and Prasetyo, 2000	32	
Lape, 2000	8	
Ames, 2001	3	4
Houston, 2001	2	

Journal of Anthropological Archaeology

Drennan and Dai, 2010	11	12
Weber and Bettinger, 2010	4	5
Potter and Chuipka, 2010	4	5
J. Peterson, 2010	3	4
Hirshman, Lovis, and Pollard, 2010	5	6
Lau, 2010	8	
Coupland, Stewart, and Patton, 2010	95	
Winterhalder et al., 2010	26	4
Nocete et al., 2010	19	8
E. E. Jones, 2010a	122	
Blair, 2010	4	5
Codding, Porcasi, and Jones, 2010	11	
Kim, 2010	4	5
Field and Lape, 2010	15	
Hart and Brumbach, 2009	25	

(continued)

Journal and source	*n* time intervals	*n* time resolutions
P. Arnold, 2009	13	5
Munson and Macri, 2009	26	
Sakaguchi, 2009	5	6
Garcea and Hildebrand, 2009	5	4
S. Jones and Pal, 2009	8	
Mantha, 2009	1	2
Small, 2009	3	4
Lyman, VanPool, and O'Brien, 2009	45	42
Mizoguchi, 2009	1	2
Huffman, 2009	35	
Hamilton and Buchanan, 2009	22	
Riel-Salvatore, Popescu, and Barton, 2008	22	
Twiss, 2008	4	5
Pitts, 2008	10	
Eriksson et al., 2008	89	
Fornander, Eriksson, and Lidén, 2008	10	
Grayson and Delpech, 2008	7	
Stahl et al., 2008	5	6
Moncel et al., 2008	11	
Vanmontfort, 2008	3	4
Mannermaa, 2008	9	
Alconini, 2008	3	4
B. Finucane, Manning, and Touré, 2008	10	
Graesch, 2007	1	2
Sara-Lafosse, 2007	26	
Langlois, 2007	4	
Littleton and Allen, 2007	22	
Sakaguchi, 2007	6	7
Dean, 2007	8	9
Delagnes et al., 2006	4	
Bicho, Haws, and Hockett, 2006	9	
E. L. Jones, 2006	6	
Shelach, 2006	19	
Jennings, 2006	1	2
Fiore and Zangrando, 2006	2	
Tafuri et al., 2006	24	
Kind, 2006	8	
Fisher, 2006	3	6
M. Betts and Friesen, 2006	4	
Bousman, 2005	13	
Yuan and Flad, 2005	2	3
Vanhaeren and d'Errico, 2005	3	
Henderson and Ostler, 2005	2	3
Mora and de la Torre, 2005	6	
Blom, 2005	7	
Mayor et al., 2005	4	
Nocete et al., 2005	29	

(continued)

(continued)

Journal and source	*n* time intervals	*n* time resolutions
M. Betts and Friesen, 2004	5	
White et al., 2004	5	
Janusek and Kolata, 2004	32	
Hardy-Smith and Edwards, 2004	2	
Hudson, 2004	2	3
Johansen, 2004	3	
Eriksson, 2004	17	
Lee, 2004	2	3
Allen, 2004	10	
Field, 2004	25	
Kohler, VanBuskirk, and Ruscavage-Barz, 2004	7	
Hastorf, 2003	3	
Alan Covey, 2003	3	
Lyman, 2003	15	
Harrison and Katzenberg, 2003	15	
van der Merwe et al., 2003	17	
Krigbaum, 2003	22	
Panja, 2003	2	3
Johnston, 2003	4	
C. Smith, 2003	40	
M. Cannon, 2003	4	5
Janetski, 2002	2	3
Weber, Link, and Katzenberg, 2002	4	5
Kim, 2001	4	6
Jennings and Craig, 2001	1	2
Schulting and Richards, 2001	13	
Schachner, 2001	11	10
Spencer and Redmond, 2001	2	3
Whitridge, 2001	2	3
M. Cannon, 2000	6	7
Porcasi, Jones, and Raab, 2000	29	3
J. Hill, 2000	9	10
Flannery and Marcus, 2000	7	9
Kuijt, 2000	4	
American Antiquity		
A. Cannon, 2000	12	
Billman, Lambert, and Leonard, 2000	14	
Stafford, Richards, and Anslinger, 2000	5	3
Maxham, 2000	4	5
Kennett and Kennett, 2000	39	
Feinman, Lightfoot, and Upham, 2000	2	3
Potter, 2000	2	3
Geib, 2000	31	5
Lewis, 2000	4	
Porcasi and Fujita, 2000	2	3
Ortman, 2000	38	11

(continued)

Journal and source	*n* time intervals	*n* time resolutions
Trubitt, 2000	3	4
Peregrine, 2001	2	3
Toll, 2001	2	3
Cameron, 2001	7	8
Mathien, 2001	6	7
Windes and McKenna, 2001	17	
Gamble, Walker, and Russell, 2001	3	4
Nelson and Hegmon, 2001	57	
Kuhn and Sempowski, 2001	6	
Richerson, Boyd, and Bettinger, 2001	16	
Wills, 2001	3	
Diaz-Granados et al., 2001	3	
Rick, Erlandson, and Vellanoweth, 2001	24	
Lovis et al., 2001	4	
Kooyman et al., 2001	3	
Kuijt, 2001	16	
Vehik, 2002	3	4
Little, 2002	14	
T. Jones et al., 2002	13	
Hildebrandt and McGuire, 2002	10	13
Gamble, 2002	5	1
Flannery, 2002	9	
Coltrain and Leavitt, 2002	32	
Kuckelman, Lightfoot, and Martin, 2002	1	
Kolb and Dixon, 2002	7	1
Gallivan, 2002	33	6
M. Beck, 2002	2	3
Lekson, 2002	3	4
Cobb and Butler, 2002	2	
Damp, Hall, and Smith, 2002	23	
Cameron, 2002	5	
Kozuch, 2002	8	7
G. Jones et al., 2003	2	3
Pauketat, 2003	4	5
Creel and Anyon, 2003	5	6
Dunham, Gold, and Hantman, 2003	6	
Ubelaker and Owsley, 2003	2	
Emerson, Hughes, Hynes, and Wisseman, 2003	7	7
Whalen and Minnis, 2003	10	
Huckell and Haynes, 2003	10	
Wheeler et al., 2003	34	
H. Jackson and Scott, 2003	1	2
Hart, Thompson, and Brumbach, 2003	2	
R. Beck, 2003	10	
W. Prentiss et al., 2003	33	6
Hodder and Cessford, 2004	17	

(continued)

(*continued*)

Journal and source	*n* time intervals	*n* time resolutions
Coulam and Schroedl, 2004	16	
Reitz, 2004	9	
DeBoer, 2004	29	
Cassidy, Raab, and Nina, 2004	3	
Henrich, 2004	5	
David and Broughton, 2004	33	
Faught, 2004	4	
Blitz and Patrick, 2004	40	
Knight, 2004	2	3
J. Hill, 2004	10	11
Van Dyke, 2004	5	
Kintigh, Donna, and Deborah, 2004	7	
Truncer, 2004	99	
Kidder, 2004	8	
Sherwood et al., 2004	4	5
Stevenson, Ihab, and Steven, 2004	8	
Schollmeyer and Turner, 2004	4	3
Deagan, 2004	3	
Deborah, 2004	5	
B. Hill et al. 2004	8	9
C. Peterson and Drennan, 2005	2	2
Wolverton, 2005	10	
Dixon, Manley, and Lee, 2005	10	
Yerkes, 2005	31	28
Diehl, 2005	3	4
Dean, 2005	7	8
T. Jones and Klar, 2005	4	5
Redmond and Tankersley, 2005	3	
Lipo, Feathers, and Dunnell, 2005	9	
Byerly et al., 2005	5	6
Saunders et al., 2005	31	
Lovis, Donahue, and Holman, 2005	3	
McGuire and Hildebrandt, 2005	8	6
Hockett, 2005	14	19
Kuzmin and Keates, 2005	33	34
A. Cannon and Yang, 2006	4	5
Kidder, 2006	145	
C. Betts, 2006	9	4
Stiger, 2006	5	
Nelson et al., 2006	1	2
Theler and Boszhardt, 2006	4	2
E. E. Jones, 2006	3	
Sassaman, Blessing, and Randall, 2006	20	
Bever, 2006	18	
Pavao-Zuckerman, 2007	2	3
Perttula and Rogers, 2007	16	
Fowles et al., 2007	1	2
J. Arnold, 2007	6	

(*continued*)

Journal and source	*n* time intervals	*n* time resolutions
Mills, 2007	5	4
Ortman, Varien, and Gripp, 2007	11	12
Varien et al., 2007	11	12
Coltrain, Janetski, and Carlyle, 2007	38	
M. Hill, 2007	37	
Cook, 2007	9	
McClure, 2007	3	4
Gibson, 2007	1	
Price, Burton, and Stoltman, 2007	2	
Hart, Brumbach, and Lusteck, 2007	37	
Braje et al., 2007	5	
Cameron and Duff, 2008	5	6
A. Prentiss et al., 2008	76	4
Geib and Jolie, 2008	12	
Ingram, 2008	5	6
Kealhofer and Grave, 2008	55	
T. Jones et al., 2008	3	4
J. Thompson, Sugiyama, and Morgan, 2008	7	4
Gremillion, Windingstad, and Sherwood, 2008	8	
Rose, 2008	3	
Cordell et al., 2008	17	
Friesen and Arnold, 2008	9	
Kohler et al., 2008	34	
Faught, 2008	59	
Hart et al., 2008	3	
Holmes et al., 2008	1	
Coupland, Clark, and Palmer, 2009	11	
V. Thompson and Turck, 2009	3	
Byers and Hill, 2009	69	
Fenner, 2009	4	
Kuijt and Goodale, 2009	8	
Robinson et al., 2009	4	
Ramenofsky, Neiman, and Pierce, 2009	4	
Abbott, 2009	6	7
Tankersley, Waters, and Stafford, 2009	5	
Wilson, 2010	2	
Galle, 2010	30	
C. Beck and Jones, 2010	39	
Boyd and Surette, 2010	3	
Perry and Jazwa, 2010	2	3
Ames, Fuld, and Davis, 2010	26	
Scarborough and Burnside, 2010	4	4
E. E. Jones, 2010b	29	
Harrower, McCorriston, and D'Andrea, 2010	22	

(*continued*)

(continued)

Journal and source	n time intervals	n time resolutions
Schachner, 2010	3	
Marquardt, 2010	3	
O'Gorman, 2010	8	
Morin, 2010	14	
Ortmann, 2010	69	
Dye, 2010	9	
Washburn, Crowe, and Ahlstrom, 2010	19	20
G. Smith, 2010	2	3
Monaghan and Peebles, 2010	10	

Citations for Appendix B

Abbott, D. R. 2009. Extensive and long-term specialization: Hohokam ceramic production in the Phoenix basin, Arizona. *American Antiquity*, 74(3):531–57.

Adams, B., and Ringer, A. 2004. New C14 dates for the Hungarian Early Upper Palaeolithic. *Current Anthropology*, 45(4):541–51.

Adams, E. 2004. Power and ritual in Neopalatial Crete: A regional comparison. *World Archaeology*, 36(1):26–42.

Adams, R. 2006. The greater Yellowstone ecosystem, soapstone bowls and the mountain Shoshone. *World Archaeology*, 38(3):528–46.

Adler, D. S., Bar Oz, G., Belfer Cohen, A., and Bar-Yosef, O. 2006. Ahead of the game: Middle and Upper Palaeolithic hunting behaviors in the southern Caucasus. *Current Anthropology*, 47(1):89–118.

Alan Covey, R. 2003. A processual study of Inka state formation. *Journal of Anthropological Archaeology*, 22(4):333–57.

Alconini, S. 2008. Dis-embedded centers and architecture of power in the fringes of the Inka Empire: New perspectives on territorial and hegemonic strategies of domination. *Journal of Anthropological Archaeology*, 27(1):63–81.

Allen, M. S. 2004. Bet-hedging strategies, agricultural change, and unpredictable environments: Historical development of dryland agriculture in Kona, Hawaii. *Journal of Anthropological Archaeology*, 23(2):196–224.

Ames, K. M. 2001. Slaves, chiefs and labour on the northern Northwest Coast. *World Archaeology*, 33(1):1–17.

Ames, K. M., Fuld, K. A., and Davis, S. 2010. Dart and arrow points on the Columbia Plateau of western North America. *American Antiquity*, 75(2):287–325.

Angelucci, D. E., Boschian, G., Fontanals, M., Pedrotti, A., and Vergès, J. M. 2009. Shepherds and karst: The use of caves and rock-shelters in the Mediterranean region during the Neolithic. *World Archaeology*, 41(2):191–214.

Arias, P. 2009. Rites in the dark? an evaluation of the current evidence for ritual areas at Magdalenian cave sites. *World Archaeology*, 41(2):262–94.

Arnold, J. E. 2007. Credit where credit is due: The history of the Chumash oceangoing plank canoe. *American Antiquity*, 72(2):196–209.

Arnold, J. E., and Bernard, J. 2005. Negotiating the coasts: Status and the evolution of boat technology in California. *World Archaeology*, 37(1):109–31.

Arnold, P. J., III. 2009. Settlement and subsistence among the Early Formative Gulf Olmec. *Journal of Anthropological Archaeology*, 28(4):397–411.

Arrizabalaga, A., Altuna, J., Areso, P., Elorza, M., García-Diez, M., Iriarte-Chiapusso, M.-J., Mariezkurrena, K., et al. 2003. The initial Upper Paleolithic in northern Iberia: New evidence from Labeko Koba. *Current Anthropology*, 44(3):413–21.

Assefa, Z., Lam, Y. M., and Mienis, H. K. 2008. Symbolic use of terrestrial gastropod Opercula during the Middle Stone Age at Porc-Epic Cave, Ethiopia. *Current Anthropology*, 49(4):746–56.

Bailey, G., and Galanidou, N. 2009. Caves, palimpsests and dwelling spaces: Examples from the Upper Palaeolithic of south-east Europe. *World Archaeology*, 41(2):215–41.

Balkansky, A. K. 2006. Surveys and Mesoamerican archaeology: The emerging macroregional paradigm. *Journal of Archaeological Research*, 14(1):53–95.

Bandy, M. S. 2005. New World settlement evidence for a two-stage Neolithic demographic transition. *Current Anthropology*, 46(S5):S109–S115.

Barber, I. 2003. Sea, land and fish: Spatial relationships and the archaeology of South Island Maori fishing. *World Archaeology*, 35(3):434–48.

———. 2010. Diffusion or innovation? Explaining lithic agronomy on the southern Polynesian margins. *World Archaeology*, 42(1):74–89.

Barham, L. S. 2002. Systematic pigment use in the middle Pleistocene of south-central Africa. *Current Anthropology*, 43(1):181–90.

Beck, C., and Jones, G. T. 2010. Clovis and Western Stemmed: Population migration and the meeting of two technologies in the intermountain West. *American Antiquity*, 75(1):81–116.

Beck, M. E. 2002. The ball-on-three-ball test for tensile strength: Refined methodology and results for three Hohokam ceramic types. *American Antiquity*, 67(3):558–69.

Beck, R. A. 2003. Consolidation and hierarchy: Chiefdom variability in the Mississippian Southeast. *American Antiquity*, 68(4):641–61.

Beekman, C. S. 2010. Recent research in western Mexican archaeology. *Journal of Archaeological Research*, 18(1):41–109.

Bell, T., and Renouf, M. A. P. 2003. Prehistoric cultures, reconstructed coasts: Maritime Archaic Indian site distribution in Newfoundland. *World Archaeology*, 35(3):350–70.

Bentley, R. A., Tayles, N., Higham, C., Macpherson, C., and Atkinson, Tim C. 2007. Shifting gender relations at Khok Phanom Di, Thailand: Isotopic evidence from the skeletons. *Current Anthropology*, 48(2):301–14.

Benz, B. F., and Long, A. 2000. Prehistoric maize evolution in the Tehuacan valley. *Current Anthropology*, 41(3):460–65.

Bettinger, R., L., Barton, L., Morgan, C., Chen, F., Wang, H., Guilderson, T. P., Ji, D., and Zhang, D. 2010. The transition to agriculture at Dadiwan, People's Republic of China. *Current Anthropology*, 51(5):703–14.

Betts, C. M. 2006. Pots and pox: The identification of protohistoric epidemics in the upper Mississippi valley. *American Antiquity*, 71(2):233–59.

Betts, M. W., and Friesen, T. M. 2004. Quantifying hunter-gatherer intensification: A zooarchaeological case study from Arctic Canada. *Journal of Anthropological Archaeology*, 23(4):357–84.

———. 2006. Declining foraging returns from an inexhaustible resource? Abundance indices and beluga whaling in the western Canadian Arctic. *Journal of Anthropological Archaeology*, 25(1):59–81.

Bever, M. R. 2006. Too little, too late? The radiocarbon chronology of Alaska and the peopling of the New World. *American Antiquity*, 71(4):595–620.

Bicho, N., Haws, J., and Hockett, B. 2006. Two sides of the same coin—rocks, bones and site function of Picareiro Cave, central Portugal. *Journal of Anthropological Archaeology*, 25(4):485–99.

Billman, B. R., Lambert, P. M., and Leonard, B. L. 2000. Cannibalism, warfare, and drought in the Mesa Verde region during the twelfth century A.D. *American Antiquity*, 65(1):145–78.

Blair, S. E. 2010. Missing the boat in lithic procurement: Watercraft and the bulk procurement of tool-stone on the maritime peninsula. *Journal of Anthropological Archaeology*, 29(1):33–46.

Blitz, J. H. 2010. New perspectives in Mississippian archaeology. *Journal of Archaeological Research*, 18(1):1–39.

Blitz, J. H., and Patrick, L. 2004. Sociopolitical implications of Mississippian mound volume. *American Antiquity*, 69(2):291–301.

Blom, D. E. 2005. Embodying borders: Human body modification and diversity in Tiwanaku society. *Journal of Anthropological Archaeology*, 24(1):1–24.

Blom, D. E., and Janusek, J. W. 2004. Making place: Humans as dedications in Tiwanaku. *World Archaeology*, 36(1):123–41.

Bocquet-Appel, J., and Naji, S. 2006. Testing the hypothesis of a worldwide Neolithic demographic transition: Corroboration from American cemeteries. *Current Anthropology*, 47(2):341–65.

Bogaard, A. 2005. "Garden agriculture" and the nature of early farming in Europe and the Near East. *World Archaeology*, 37(2):177–96.

Borrero, L. A., and Barberena, R. 2006. Hunter-gatherer home ranges and marine resources: An archaeological case from southern Patagonia. *Current Anthropology*, 47(5):855–67.

Bousman, C. B. 2005. Coping with risk: Later Stone Age technological strategies at Blydefontein Rock Shelter, South Africa. *Journal of Anthropological Archaeology*, 24(3):193–226.

Boyd, M., and Surette, C. 2010. Northernmost precontact maize in North America. *American Antiquity*, 75(1):117–33.

Braemer, F., Genequand, D., Maridat, C. D., Blanc, P. M., Dentzer, J. M., Gazagne, D., and Wech, P. 2009. Long-term management of water in the central Levant: The Hawran case (Syria). *World Archaeology*, 41(1):36–57.

Braje, T. J., Kennett, D. J., Erlandson, J. M., and Culleton, B. J. 2007. Human impacts on near-shore shellfish taxa: A 7,000 year record from Santa Rosa Island, California. *American Antiquity*, 72(4):735–56.

Brantingham, P. J., Krivoshapkin, A. I., Jinzeng, L., and Tserendagva, Y. 2001. The initial Upper Paleolithic in Northeast Asia. *Current Anthropology*, 42(5):735–46.

Brantingham, P. J., and Xing, G. 2006. Peopling of the northern Tibetan Plateau. *World Archaeology*, 38(3):387–414.

Bright, J., Ugan, A., and Hunsaker, L. 2002. The effect of handling time on subsistence technology. *World Archaeology*, 34(1):164–81.

Bulbeck, F. D., and Prasetyo, B. 2000. Two millennia of socio-cultural development in Luwu, South Sulawesi, Indonesia. *World Archaeology*, 32(1):121–37.

Byerly, R. M., Cooper, J. R., Meltzer, D. J., Hill, M. E., and LaBelle, J. M. 2005. On Bonfire Shelter (Texas) as a Paleoindian bison jump: An assessment using GIS and zooarchaeology. *American Antiquity*, 70(4):595–629.

Byers, D. A., and Hill, B. L. 2009. Pronghorn dental age profiles and Holocene hunting strategies at Hogup Cave, Utah. *American Antiquity*, 74(2):299–321.

Byrd, B. F. 2005. Reassessing the emergence of village life in the Near East. *Journal of Archaeological Research*, 13(3):231–90.

Cameron, C. M. 2001. Pink chert, projectile points, and the Chacoan regional system. *American Antiquity*, 66(1):79–101.

———. 2002. Sacred earthen architecture in the northern Southwest: The Bluff Great House berm. *American Antiquity*, 67(4):677–95.

Cameron, C. M., and Duff, A. I. 2008. History and process in village formation: Context and contrasts from the northern Southwest. *American Antiquity*, 73(1):29–57.

Cannon, A. 2000. Settlement and sea-levels on the central coast of British Columbia: Evidence from shell midden cores. *American Antiquity*, 65(1):67–77.

Cannon, A., and Yang, D. Y. 2006. Early storage and sedentism on the Pacific Northwest Coast: Ancient DNA analysis of salmon remains from Namu, British Columbia. *American Antiquity*, 71(1):123–40.

Cannon, M. D. 2000. Large mammal relative abundance in Pithouse and Pueblo period archaeofaunas from southwestern New Mexico: Resource depression among the Mimbres-Mogollon? *Journal of Anthropological Archaeology*, 19(3):317–47.

———. 2003. A model of central place forager prey choice and an application to faunal remains from the Mimbres valley, New Mexico. *Journal of Anthropological Archaeology*, 22(1):1–25.

Cassidy, J., Raab, L. M., and Nina, A. K. 2004. Boats, bones, and biface bias: The early Holocene mariners of Eel Point, San Clemente Island, California. *American Antiquity*, 69(1):109–30.

Chatters, J. C., and Prentiss, W. C. 2005. A Darwinian macro-evolutionary perspective on the development of hunter-gatherer systems in northwestern North America. *World Archaeology*, 37(1):46–65.

Cobb, C. R., and Butler, B. M. 2002. The vacant quarter revisited: Late Mississippian abandonment of the lower Ohio valley. *American Antiquity*, 67(4):625–41.

Codding, B. F., Porcasi, J. F., and Jones, T. L. 2010. Explaining prehistoric variation in the abundance of large prey: A zooarchaeological analysis of deer and rabbit hunting along the Pecho Coast of central California. *Journal of Anthropological Archaeology*, 29(1):47–61.

Coltrain, J. B., Hayes, M. G., and O'Rourke, D. H. 2006. Hrdlička's Aleutian population-replacement hypothesis: A radiometric evaluation. *Current Anthropology*, 47(3):537–48.

Coltrain, J. B., Janetski, J. C., and Carlyle, S. W. 2007. The stable- and radio-isotope chemistry of western basketmaker burials: Implications for early Puebloan diets and origins. *American Antiquity*, 72(2):301–21.

Coltrain, J. B., and Leavitt, S. W. 2002. Climate and diet in Fremont prehistory: Economic variability and abandonment of maize agriculture in the Great Salt Lake Basin. *American Antiquity*, 67(3):453–85.

Cook, R. A. 2007. Single component sites with long sequences of radiocarbon dates: The Sunwatch site and Middle Fort ancient village growth. *American Antiquity*, 72(3):439–60.

Cooke, R. 2005. Prehistory of Native Americans on the Central American land bridge: Colonization, dispersal, and divergence. *Journal of Archaeological Research*, 13(2):129–87.

Cordell, L. S., Toll, H. W., Toll, M. S., and Windes, T. C. 2008. Archaeological corn from Pueblo Bonito, Chaco Canyon, New Mexico: Dates, contexts, sources. *American Antiquity*, 73(3):491–511.

Coulam, N. J., and Schroedl, A. R. 2004. Late Archaic totemism in the greater American Southwest. *American Antiquity*, 69(1):41–62.

Coupland, G., Clark, T., and Palmer, A. 2009. Hierarchy, communalism, and the spatial order of Northwest Coast plank houses: A comparative study. *American Antiquity*, 74(1):77–106.

Coupland, G., Stewart, K., and Patton, K. 2010. Do you never get tired of salmon? Evidence for extreme salmon specialization at Prince Rupert Harbour, British Columbia. *Journal of Anthropological Archaeology*, 29(2):189–207.

Creel, D., and Anyon, R. 2003. New interpretations of Mimbres public architecture and space: Implications for cultural change. *American Antiquity*, 68(1):67–92.

Damp, J. E., Hall, S. A., and Smith, S. J. 2002. Early irrigation on the Colorado Plateau near Zuni Pueblo, New Mexico. *American Antiquity*, 67(4):665–76.

David, A. B., and Broughton, J. M. 2004. Holocene environmental change, artiodactyl abundances, and human hunting strategies in the Great Basin. *American Antiquity*, 69(2):235–55.

Deagan, K. 2004. Reconsidering Taíno social dynamics after Spanish conquest: Gender and class in culture contact studies. *American Antiquity*, 69(4):597–626.

Dean, R. M. 2005. Site-use intensity, cultural modification of the environment, and the development of agricultural communities in southern Arizona. *American Antiquity*, 70(3):403–31.

———. 2007. Hunting intensification and the Hohokam "collapse." *Journal of Anthropological Archaeology*, 26(1):109–32.

de Beaune, S. A. 2004. The invention of technology: Prehistory and cognition. *Current Anthropology*, 45(2):139–62.

DeBoer, W. R. 2004. Little Bighorn on the Scioto: The Rocky Mountain connection to Ohio Hopewell. *American Antiquity*, 69(1):85–107.

Deborah, A. K. 2004. Reevaluating late prehistoric coastal subsistence and settlement strategies: New data from Grove's Creek Site, Skidaway Island, Georgia. *American Antiquity*, 69(4):671–88.

Delagnes, A., Lenoble, A., Harmand, S., Brugal, J.-P., Prat, S., Tiercelin, J.-J., and Roche, H. 2006. Interpreting pachyderm single carcass sites in the African lower and early middle Pleistocene record: A multidisciplinary approach to the site of Nadung'a 4 (Kenya). *Journal of Anthropological Archaeology*, 25(4):448–65.

de la Torre, I. 2004. Omo revisited: Evaluating the technological skills of Pliocene hominids. *Current Anthropology*, 45(4):439–65.

Denham, T. 2005. Envisaging early agriculture in the Highlands of New Guinea: Landscapes, plants and practices. *World Archaeology*, 37(2):290–306.

Diaz-Granados, C., Rowe, M. W., Hyman, M., Duncan, J. R., and Southon, J. R. 2001. AMS radiocarbon dates for charcoal from three Missouri pictographs and their associated iconography. *American Antiquity*, 66(3):481–92.

Diehl, M. W. 2005. Morphological observations on recently recovered early agricultural period maize cob fragments from southern Arizona. *American Antiquity*, 70(2):361–75.

Dixon, E. J., Manley, W. F., and Lee, C. M. 2005. The emerging archaeology of glaciers and ice patches: Examples from Alaska's Wrangell–St. Elias National Park and Preserve. *American Antiquity*, 70(1):129–43.

Dizon, E., Détroit, F., Sémah, F., Falguères, C., Hameau, S., Ronquillo, W., and Cabanis, E. 2002. Notes on the morphology and age of the Tabon Cave fossil *Homo sapiens*. *Current Anthropology*, 43(4):660–66.

Drennan, R. D., and Dai, X. 2010. Chiefdoms and states in the Yuncheng basin and the Chifeng region: A comparative analysis of settlement systems in North China. *Journal of Anthropological Archaeology*, 29(4):455–68.

Dunham, G. H., Gold, D. L., and Hantman, J. L. 2003. Collective burial in late prehistoric Virginia: Excavation and analysis of the Rapidan Mound. *American Antiquity*, 68(1):109–28.

Dye, T. S. 2010. Social transformation in Old Hawai'i: A bottom-up approach. *American Antiquity*, 75(4):727–41.

Efstratiou, N., Biagi, P., Elefanti, P., Karkanas, P., and Ntinou, M. 2006. Prehistoric exploitation of Grevena Highland zones: Hunters and herders along the Pindus Chain of Western Macedonia (Greece). *World Archaeology*, 38(3):415–35.

Emerson, T. E., Hughes, R. E., Hynes, M. R., and Wisseman, S. U. 2003. The sourcing and interpretation of Cahokia-style figurines in the trans-Mississippi South and Southeast. *American Antiquity*, 68(2):287–313.

Eriksson, G. 2004. Part-time farmers or hard-core sealers? Västerbjers studied by means of stable isotope analysis. *Journal of Anthropological Archaeology*, 23(2):135–62.

Eriksson, G., Linderholm, A., Fornander, E., Kanstrup, M., Schoultz, P., Olofsson, H., and Lidén, K. 2008. Same island, different diet: Cultural evolution of food practice on Öland, Sweden, from the Mesolithic to the Roman period. *Journal of Anthropological Archaeology*, 27(4):520–43.

Erlandson, J. M. 2001. The archaeology of aquatic adaptations: Paradigms for a new millennium. *Journal of Archaeological Research*, 9(4):287–350.

Fairbairn, A. 2005a. An archaeobotanical perspective on Holocene plant-use practices in lowland northern New Guinea. *World Archaeology*, 37(4):487–502.

———. 2005b. A history of agricultural production at Neolithic Çatalhöyük East, Turkey. *World Archaeology*, 37(2):197–210.

Faught, M. K. 2004. The underwater archaeology of paleolandscapes, Apalachee Bay, Florida. *American Antiquity*, 69(2):275–89.

———. 2008. Archaeological roots of human diversity in the New World: A compilation of accurate and precise radiocarbon ages from earliest sites. *American Antiquity*, 73(4):670–98.

Feinman, G. M., Lightfoot, K. G., and Upham, S. 2000. Political hierarchies and organizational strategies in the Puebloan Southwest. *American Antiquity*, 65(3):449–70.

Fenner, J. N. 2009. Occasional hunts or mass kills? Investigating the origins of archaeological pronghorn bonebeds in southwest Wyoming. *American Antiquity*, 74(2):323–50.

Fiedel, S. J. 2000. The peopling of the New World: Present evidence, new theories, and future directions. *Journal of Archaeological Research*, 8(1):39–103.

———. 2005. Man's best friend—mammoth's worst enemy? A speculative essay on the role of dogs in Paleoindian colonization and megafaunal extinction. *World Archaeology*, 37(1): 11–25.

Field, J. S. 2004. Environmental and climatic considerations: A hypothesis for conflict and the emergence of social complexity in Fijian prehistory. *Journal of Anthropological Archaeology*, 23(1):79–99.

Field, J. S., and Lape, P. V. 2010. Paleoclimates and the emergence of fortifications in the tropical Pacific Islands. *Journal of Anthropological Archaeology*, 29(1):113–24.

Finucane, B., Manning, K., and Touré, M. 2008. Late Stone Age subsistence in the Tilemsi valley, Mali: Stable isotope analysis of human and animal remains from the site of Karkarichinkat Nord (KN05) and Karkarichinkat Sud (KS05). *Journal of Anthropological Archaeology*, 27(1):82–92.

Finucane, B. C. 2009. Maize and sociopolitical complexity in the Ayacucho valley, Peru. *Current Anthropology*, 50(4):535–45.

Fiore, D., and Zangrando, A. F. J. 2006. Painted fish, eaten fish: Artistic and archaeofaunal representations in Tierra del Fuego, southern South America. *Journal of Anthropological Archaeology*, 25(3):371–89.

Fisher, L. E. 2006. Blades and microliths: Changing contexts of tool production from Magdalenian to early Mesolithic in southern Germany. *Journal of Anthropological Archaeology*, 25(2):226–38.

Flad, R. K. 2008. Divination and power: A multiregional view of the development of oracle bone divination in early China. *Current Anthropology*, 49(3):403–37.

Flannery, K. V. 2002. The origins of the village revisited: From nuclear to extended households. *American Antiquity*, 67(3):417–33.

Flannery, K. V., and Marcus, J. 2000. Formative Mexican chiefdoms and the myth of the "mother culture." *Journal of Anthropological Archaeology*, 19(1):1–37.

Fornander, E., Eriksson, G., and Lidén, K. 2008. Wild at heart: Approaching pitted ware identity, economy and cosmology through stable isotopes in skeletal material from the Neolithic site Korsnäs in eastern central Sweden. *Journal of Anthropological Archaeology*, 27(3): 281–97.

Fowles, S. M., Minc, L., Duwe, S., and Hill, D. V. 2007. Clay, conflict, and village aggregation: Compositional analyses of pre-Classic pottery from Taos, New Mexico. *American Antiquity*, 72(1):125–52.

Frachetti, M. D., Benecke, N., Mar'yashev, A. N., and Doumani, P. N. 2010. Eurasian pastoralists and their shifting regional interactions at the steppe margin: Settlement history at Mukri, Kazakhstan. *World Archaeology*, 42(4):622–46.

Friesen, T. M. 2004. Contemporaneity of Dorset and Thule cultures in the North American Arctic: New radiocarbon dates from Victoria Island, Nunavut. *Current Anthropology*, 45(5):685–91.

Friesen, T. M., and Arnold, C. D. 2008. The timing of the Thule migration: New dates from the western Canadian Arctic. *American Antiquity*, 73(3):527–38.

Frink, L. 2007. Storage and status in precolonial and colonial coastal western Alaska. *Current Anthropology*, 48(3):349–74.

Fuller, D. Q., and Qin, L. 2009. Water management and labour in the origins and dispersal of Asian rice. *World Archaeology*, 41(1):88–111.

Galle, J. E. 2010. Costly signaling and gendered social strategies among slaves in the eighteenth-century Chesapeake: An archaeological perspective. *American Antiquity*, 75(1):19–43.

Gallivan, M. D. 2002. Measuring sedentariness and settlement population: Accumulations research in the Middle Atlantic region. *American Antiquity*, 67(3):535–57.

Gamble, L. H. 2002. Archaeological evidence for the origin of the plank canoe in North America. *American Antiquity*, 67(2):301–15.

Gamble, L. H., Walker, P. L., and Russell, G. S. 2001. An integrative approach to mortuary analysis: Social and symbolic dimensions of Chumash burial practices. *American Antiquity*, 66(2):185–212.

Garcea, E. A. A. 2006. Semi-permanent foragers in semi-arid environments of North Africa. *World Archaeology*, 38(2):197–219.

Garcea, E. A. A., and Hildebrand, E. A. 2009. Shifting social networks along the Nile: Middle Holocene ceramic assemblages from Sai Island, Sudan. *Journal of Anthropological Archaeology*, 28(3):304–22.

Garcia Guix, E., Richards, M. P., and Subir, M. E. 2006. Palaeodiets of humans and fauna at the Spanish Mesolithic site of El Collado. *Current Anthropology*, 47(3):549–56.

Geib, P. R. 2000. Sandal types and Archaic prehistory on the Colorado Plateau. *American Antiquity*, 65(3):509–24.

Geib, P. R., and Jolie, E. A. 2008. The role of basketry in early Holocene small seed exploitation: Implications of a ca. 9,000 year-old basket from Cowboy Cave, Utah. *American Antiquity*, 73(1):83–102.

Gibaja Bao, J. F.. 2004. Neolithic communities of the northeastern Iberian Peninsula: Burials, grave goods, and lithic tools. *Current Anthropology*, 45(5):679–85.

Gibson, J. L. 2006. Navels of the earth: Sedentism in early mound-building cultures in the lower Mississippi valley. *World Archaeology*, 38(2):311–29.

———. 2007. "Formed from the earth at that place": The material side of community at Poverty Point. *American Antiquity*, 72(3):509–23.

Gil, A. F. 2003. *Zea mays* on the South American periphery: Chronology and dietary importance. *Current Anthropology*, 44(2):295–300.

Golovanova, L. V., Doronichev, V. B., Cleghorn, N. E., Koulkova, M. A., Sapelko, T. V., and Shackley, M. S. 2010. Significance of ecological factors in the Middle to Upper Paleolithic transition. *Current Anthropology*, 51(5):655–91.

Graesch, A. P. 2007. Modeling ground slate knife production and implications for the study of household labor contributions to salmon fishing on the Pacific Northwest Coast. *Journal of Anthropological Archaeology*, 26(4):576–606.

Grayson, D. K., and Delpech, F. 2008. The large mammals of Roc de Combe (Lot, France): The Châtelperronian and Aurignacian assemblages. *Journal of Anthropological Archaeology*, 27(3):338–62.

Gremillion, K. J., Windingstad, J., and Sherwood, S. C. 2008. Forest opening, habitat use, and food production on the Cumberland Plateau, Kentucky: Adaptive flexibility in marginal settings. *American Antiquity*, 73(3):387–411.

Haas, J., and Creamer, W. 2006. Crucible of Andean civilization: The Peruvian coast from 3000 to 1800 BC. *Current Anthropology*, 47(5):745–75.

Hamilton, M. J., and Buchanan, B. 2009. The accumulation of stochastic copying errors causes drift in culturally transmitted technologies: Quantifying Clovis evolutionary dynamics. *Journal of Anthropological Archaeology*, 28(1):55–69.

Hardy-Smith, T., and Edwards, P. C. 2004. The garbage crisis in prehistory: Artefact discard patterns at the Early Natufian site of Wadi Hammeh 27 and the origins of household refuse disposal strategies. *Journal of Anthropological Archaeology*, 23(3):253–89.

Harrison, R. G., and Katzenberg, M. A. 2003. Paleodiet studies using stable carbon isotopes from bone apatite and collagen: Examples from southern Ontario and San Nicolas Island, California. *Journal of Anthropological Archaeology*, 22(3):227–44.

Harrower, M. J. 2008. Hydrology, ideology, and the origins of irrigation in ancient southwest Arabia. *Current Anthropology*, 49(3):497–510.

Harrower, M. J., McCorriston, J., and D'Andrea, A. C. 2010. General/specific, local/global: Comparing the beginning of agriculture in the Horn of Africa (Ethiopia/Eritrea) and southwest Arabia (Yemen). *American Antiquity*, 75(3):452–72.

Hart, J. P., and Brumbach, H. J. 2009. On pottery change and northern Iroquoian origins: An assessment from the Finger Lakes region of central New York. *Journal of Anthropological Archaeology*, 28(4):367–81.

Hart, J. P., Brumbach, H. J., and Lusteck, R. 2007. Extending the phytolith evidence for early maize (*Zea mays* ssp. *mays*) and Squash (*Cucurbita* sp.) in Central New York. *American Antiquity*, 72(3):563–83.

Hart, J. P., Reber, E. A., Thompson, R. G., and Lusteck, R. 2008. Taking variation seriously: Testing the steatite mast-processing hypothesis with microbotanical data from the Hunter's Home site, New York. *American Antiquity*, 73(4):729–41.

Hart, J. P., Thompson, R. G., and Brumbach, H. J. 2003. Phytolith evidence for early maize (*Zea mays*) in the northern Finger Lakes region of New York. *American Antiquity*, 68(4): 619–40.

Hastorf, C. A. 2003. Community with the ancestors: Ceremonies and social memory in the Middle Formative at Chiripa, Bolivia. *Journal of Anthropological Archaeology*, 22(4):305–32.

Hegmon, M. 2002. Recent issues in the archaeology of the Mimbres region of the North American Southwest. *Journal of Archaeological Research*, 10(4):307–57.

Henderson, H., and Ostler, N. 2005. Muisca settlement organization and chiefly authority at Suta, Valle de Leyva, Colombia: A critical appraisal of native concepts of house for studies of complex societies. *Journal of Anthropological Archaeology*, 24(2):148–78.

Henrich, J. 2004. Demography and cultural evolution: How adaptive cultural processes can produce maladaptive losses; The Tasmanian case. *American Antiquity*, 69(2):197–214.

Hildebrandt, W. R., and McGuire, K. R. 2002. The ascendance of hunting during the California Middle Archaic: An evolutionary perspective. *American Antiquity*, 67(2):231–56.

Hill, B. H., Clark, J. J., Doelle, W. H., and Lyons, P. D. 2004. Prehistoric demography in the Southwest: Migration, coalescence, and Hohokam population decline. *American Antiquity*, 69(4):689–716.

Hill, J. B. 2000. Decision making at the margins: Settlement trends, temporal scale, and ecology in the Wadi Al-Hasa, west-central Jordan. *Journal of Anthropological Archaeology*, 19(2):221–41.

———. 2004. Land use and an archaeological perspective on socio-natural studies in the Wadi Al-Hasa, west-central Jordan. *American Antiquity*, 69(3):389–412.

Hill, M. E. 2007. A moveable feast: Variation in faunal resource use among central and western North American Paleoindian sites. *American Antiquity*, 72(3):417–38.

Hirshman, A. J., Lovis, W. A., and Pollard, H. P. 2010. Specialization of ceramic production: A sherd assemblage based analytic perspective. *Journal of Anthropological Archaeology*, 29(3):265–77.

Hockett, B. 2005. Middle and late Holocene hunting in the Great Basin: A critical review of the debate and future prospects. *American Antiquity*, 70(4):713–31.

Hodder, I., and Cessford, C. 2004. Daily practice and social memory at Çatalhöyük. *American Antiquity*, 69(1):17–40.

Holmes, C. E., Potter, B. A., Reuther, J. D., Mason, O. K., Thorson, R. M., and Bowers, P. M. 2008. Geological and cultural context of the Nogahabara I site. *American Antiquity*, 73(4):781–90.

Hoopes, J. W. 2005. The emergence of social complexity in the Chibchan world of southern Central America and northern Colombia, AD 300–600. *Journal of Archaeological Research*, 13(1):1–47.

Houston, S. D. 2001. Decorous bodies and disordered passions: Representations of emotion among the Classic Maya. *World Archaeology*, 33(2):206–19.

Huckell, B. B., and Haynes, C. V. 2003. The Ventana Complex: New dates and new ideas on its place in early Holocene western prehistory. *American Antiquity*, 68(2):353–71.

Hudson, M. J. 2004. The Perverse realities of change: World system incorporation and the Okhotsk culture of Hokkaido. *Journal of Anthropological Archaeology*, 23(3):290–308.

Huffman, T. N. 2009. Mapungubwe and Great Zimbabwe: The origin and spread of social complexity in southern Africa. *Journal of Anthropological Archaeology*, 28(1):37–54.

Ingram, S. E. 2008. Streamflow and population change in the lower Salt River valley of central Arizona, ca. A.D. 775 to 1450. *American Antiquity*, 73(1):136–65.

Iriarte, J. 2006. Landscape transformation, mounded villages and adopted cultigens: The rise of early formative communities in south-eastern Uruguay. *World Archaeology*, 38(4):644–63.

Jackson, D., Méndez, C., Seguel, R., Maldonado, A., and Vargas, G. 2007. Initial occupation of the Pacific Coast of Chile during late Pleistocene times. *Current Anthropology*, 48(5):725–31.

Jackson, H. E., and Scott, S. L. 2003. Patterns of elite faunal utilization at Moundville, Alabama. *American Antiquity*, 68(3):552–72.

James, H. V. A., and Petraglia, M. D. 2005. Modern human origins and the evolution of behavior in the later Pleistocene record of South Asia. *Current Anthropology*, 46(S5):S3–S27.

Janetski, J. C. 2002. Trade in Fremont society: Contexts and contrasts. *Journal of Anthropological Archaeology*, 21(3):344–70.

Janusek, J. W. 2004. Tiwanaku and its precursors: Recent research and emerging perspectives. *Journal of Archaeological Research*, 12(2):121–83.

———. 2006. The changing "nature" of Tiwanaku religion and the rise of an Andean state. *World Archaeology*, 38(3):469–92.

Janusek, J. W., and Kolata, A. L. 2004. Top-down or bottom-up: Rural settlement and raised field agriculture in the Lake Titicaca basin, Bolivia. *Journal of Anthropological Archaeology*, 23(4):404–30.

Jennings, J. 2006. Core, peripheries, and regional realities in Middle Horizon Peru. *Journal of Anthropological Archaeology*, 25(3):346–70.

Jennings, J., and Craig, N. 2001. Politywide analysis and imperial political economy: The relationship between valley political complexity and administrative centers in the Wari Empire of the central Andes. *Journal of Anthropological Archaeology*, 20(4):479–502.

Johansen, P. G. 2004. Landscape, monumental architecture, and ritual: A reconsideration of the South Indian ashmounds. *Journal of Anthropological Archaeology*, 23(3):309–30.

Johnston, K. J. 2003. The intensification of pre-industrial cereal agriculture in the tropics: Boserup, cultivation lengthening, and the Classic Maya. *Journal of Anthropological Archaeology*, 22(2):126–61.

Jones, E. E. 2006. Using viewshed analysis to explore settlement choice: A case study of the Onondaga Iroquois. *American Antiquity*, 71(3):523–38.

———. 2010a. An analysis of factors influencing sixteenth and seventeenth century Haudenosaunee (Iroquois) settlement locations. *Journal of Anthropological Archaeology*, 29(1):1–14.

———. 2010b. Population history of the Onondaga and Oneida Iroquois, A.D. 1500–1700. *American Antiquity*, 75(2):387–407.

Jones, E. L. 2006. Prey choice, mass collecting, and the wild European rabbit (*Oryctolagus cuniculus*). *Journal of Anthropological Archaeology*, 25(3):275–89.

Jones, G. T., Beck, C., Jones, E. E., and Hughes, R. E. 2003. Lithic source use and Paleoarchaic foraging territories in the Great Basin. *American Antiquity*, 68(1):5–38.

Jones, S. C., and Pal, J. N. 2009. The Palaeolithic of the Middle Son valley, north-central India: Changes in hominin lithic technology and behaviour during the Upper Pleistocene. *Journal of Anthropological Archaeology*, 28(3):323–41.

Jones, T. L., Fitzgerald, R. T., Kennett, D. J., Miksicek, C. H., Fagan, J. L., Sharp, J., and Erlandson, J. M. 2002. The Cross Creek site (CA-SLO-1797) and its implications for New World colonization. *American Antiquity*, 67(2):213–30.

Jones, T. L., and Klar, K. A. 2005. Diffusionism reconsidered: Linguistic and archaeological evidence for prehistoric Polynesian contact with southern California. *American Antiquity*, 70(3):457–84.

Jones, T. L., Porcasi, J. F., Gaeta, J. W., and Codding, B. F. 2008. The Diablo Canyon fauna: A coarse-grained record of trans-Holocene foraging from the central California mainland coast. *American Antiquity*, 73(2):289–316.

Kansa, S. W., Kennedy, A., Campbell, S., and Carter, E. 2009. Resource exploitation at late Neolithic Domuztepe: Faunal and botanical evidence. *Current Anthropology*, 50(6):897–914.

Kealhofer, L., and Grave, P. 2008. Land use, political complexity, and urbanism in mainland Southeast Asia. *American Antiquity*, 73(2):200–225.

Keegan, W. F. 2000. West Indian archaeology, 3: Ceramic age. *Journal of Archaeological Research*, 8(2):135–67.

Keita, S. O. Y. 2003. A study of vault porosities in early Upper Egypt from the Badarian through Dynasty I. *World Archaeology*, 35(2):210–22.

Kennett, D. J., and Kennett, J. P. 2000. Competitive and cooperative responses to climatic instability in coastal southern California. *American Antiquity*, 65(2):379–95.

Kidder, T. R. 2004. Plazas as architecture: An example from the Raffman site, northeast Louisiana. *American Antiquity*, 69(3):514–32.

———. 2006. Climate change and the Archaic to Woodland transition (3000–2500 Cal B.P.) in the Mississippi River basin. *American Antiquity*, 71(2):195–231.

Kim, J. 2001. Elite strategies and the spread of technological innovation: The spread of iron in the Bronze Age societies of Denmark and southern Korea. *Journal of Anthropological Archaeology*, 20(4):442–78.

———. 2010. Opportunistic versus target mode: Prey choice changes in central-western Korean prehistory. *Journal of Anthropological Archaeology*, 29(1):80–93.

Kind, C.-J. 2006. Transport of lithic raw material in the Mesolithic of southwest Germany. *Journal of Anthropological Archaeology*, 25(2):213–25.

King, A. 2003. Over a century of explorations at Etowah. *Journal of Archaeological Research*, 11(4):279–306.

Kintigh, K. W., Donna, M. G., and Deborah, L. H. 2004. Long-term settlement history and the emergence of towns in the Zuni area. *American Antiquity*, 69(3):432–56.

Knight, J. V. J. 2004). Characterizing elite midden deposits at Moundville. *American Antiquity*, 69(2):304–21.

Kohler, T. A., Glaude, M. P., Bocquet-Appel, J.-P., and Kemp, B. M. 2008. The Neolithic demographic transition in the U.S. Southwest. *American Antiquity*, 73(4):645–69.

Kohler, T. A., and Turner, K. K. 2006. Raiding for women in the pre-Hispanic northern Pueblo Southwest? A pilot examination. *Current Anthropology*, 47(6):1035–1045.

Kohler, T. A., VanBuskirk, S., and Ruscavage-Barz, S. 2004. Vessels and villages: evidence for conformist transmission in early village aggregations on the Pajarito Plateau, New Mexico. *Journal of Anthropological Archaeology*, 23(1):100–118.

Kolb, M. J. 2006. The origins of monumental architecture in ancient Hawai'i. *Current Anthropology*, 47(4):657–65.

Kolb, M. J., and Dixon, B. 2002. Landscapes of war: Rules and conventions of conflict in ancient Hawai'i (and elsewhere). *American Antiquity*, 67(3):514–34.

Kooyman, B., Newman, M. E., Cluney, C., Lobb, M., Tolman, S., McNeil, P., and Hills, L. V. 2001. Identification of horse exploitation by Clovis hunters based on protein analysis. *American Antiquity*, 66(4):686–91.

Kozuch, L. 2002. Olivella beads from Spiro and the Plains. *American Antiquity*, 67(4):697–709.

Krigbaum, J. 2003. Neolithic subsistence patterns in northern Borneo reconstructed with stable carbon isotopes of enamel. *Journal of Anthropological Archaeology*, 22(3):292–304.

Kuckelman, K. A., Lightfoot, R. R., and Martin, D. L. 2002. The bioarchaeology and taphonomy of violence at Castle Rock and Sand Canyon Pueblos, southwestern Colorado. *American Antiquity*, 67(3):486–513.

Kuhn, R. D., and Sempowski, M. L. 2001. A new approach to dating the League of the Iroquois. *American Antiquity*, 66(2):301–14.

Kuijt, I. 2000. People and space in early agricultural villages: Exploring daily lives, community size, and architecture in the late Pre-pottery Neolithic. *Journal of Anthropological Archaeology*, 19(1):75–102.

———. 2001. Reconsidering the cause of cultural collapse in the Lillooet area of British Columbia, Canada: A geoarchaeological perspective. *American Antiquity*, 66(4):692–703.

Kuijt, I., and Goodale, N. 2009. Daily practice and the organization of space at the dawn of agriculture: A case study from the Near East. *American Antiquity*, 74(3):403–22.

Kuzmin, Y. V. 2008. Siberia at the last glacial maximum: Environment and archaeology. *Journal of Archaeological Research*, 16(2):163–223.

Kuzmin, Y. V., and Keates, S. G. 2005. Dates are not just data: Paleolithic settlement patterns in Siberia derived from radiocarbon records. *American Antiquity*, 70(4):773–89.

Ladefoged, T. N., and Graves, M. W. 2008. Variable development of dryland agriculture in Hawai'i: A fine-grained chronology from the Kohala field system, Hawai'i Island. *Current Anthropology*, 49(5):771–802.

Lahiri, N., and Bacus, E. A. 2004. Exploring the archaeology of Hinduism. *World Archaeology*, 36(3):313–25.

Lambert, P. M. 2002. The archaeology of war: A North American perspective. *Journal of Archaeological Research*, 10(3):207–41.

Langlois, O. 2007. Intrasite features distribution as a source of social information: The case of Djaba-Hosséré (northern Cameroon). *Journal of Anthropological Archaeology*, 26(2):172–97.

Lape, P. V. 2000. Political dynamics and religious change in the late pre-colonial Banda Islands, eastern Indonesia. *World Archaeology*, 32(1):138–55.

Latinis, D. K. 2000. The development of subsistence system models for island Southeast Asia and near Oceania: The nature and role of arboriculture and arboreal-based economies. *World Archaeology*, 32(1):41–67.

Lau, G. F. 2010. House forms and Recuay culture: Residential compounds at Yayno (Ancash, Peru), a fortified hilltop town, AD 400–800. *Journal of Anthropological Archaeology*, 29(3):327–51.

Lee, Y. K. 2004. Control strategies and polity competition in the lower Yi-Luo valley, North China. *Journal of Anthropological Archaeology*, 23(2):172–95.

Lekson, S. H. 2002. War in the Southwest, war in the world. *American Antiquity*, 67(4):607–24.

Lewis, R. B. 2000. Sea-level rise and subsidence effects on Gulf Coast archaeological site distributions. *American Antiquity*, 65(3):525–41.

Lipo, C. P., Feathers, J. K., and Dunnell, R. C. 2005. Temporal data requirements, luminescence dates, and the resolution of chronological structure of late prehistoric deposits in the central Mississippi River valley. *American Antiquity*, 70(3):527–44.

Little, E. A. 2002. Kautantouwit's legacy: Calibrated dates on prehistoric maize in New England. *American Antiquity*, 67(1):109–18.

Littleton, J., and Allen, H. 2007. Hunter-gatherer burials and the creation of persistent places in southeastern Australia. *Journal of Anthropological Archaeology*, 26(2):283–98.

Lovis, W. A., Donahue, R. E., and Holman, M. B. 2005. Long-distance logistic mobility as an organizing principle among northern hunter-gatherers: A Great Lakes middle Holocene settlement system. *American Antiquity*, 70(4):669–93.

Lovis, W. A., Egan-Bruhy, K. C., Smith, B. A., and Monaghan, G. W. 2001. Wetlands and emergent horticultural economies in the upper Great Lakes: A new perspective from the Schultz site. *American Antiquity*, 66(4):615–32.

Lucero, L. J. 2003. The politics of ritual: The emergence of Classic Maya rulers. *Current Anthropology*, 44(4):523–58.

Lyman, R. L. 2003. Pinniped behavior, foraging theory, and the depression of metapopulations and nondepression of a local population on the southern Northwest Coast of North America. *Journal of Anthropological Archaeology*, 22(4):376–88.

Lyman, R. L., VanPool, T. L., and O'Brien, M. J. 2009. The diversity of North American projectile-point types, before and after the bow and arrow. *Journal of Anthropological Archaeology*, 28(1):1–13.

MacKinnon, M. 2010. "Sick as a dog": Zooarchaeological evidence for pet dog health and welfare in the Roman world. *World Archaeology*, 42(2):290–309.

Mannermaa, K. 2008. Birds and burials at Ajvide (Gotland, Sweden) and Zvejnieki (Latvia) about 8000–3900 BP. *Journal of Anthropological Archaeology*, 27(2):201–25.

Mantha, A. 2009. Territoriality, social boundaries and ancestor veneration in the central Andes of Peru. *Journal of Anthropological Archaeology*, 28(2):158–76.

Marquardt, W. H. 2010. Shell mounds in the southeast: Middens, monuments, temple mounds, rings, or works? *American Antiquity*, 75(3):551–70.

Mathien, F. J. 2001. The organization of turquoise production and consumption by the prehistoric Chacoans. *American Antiquity*, 66(1):103–18.

Matsui, A., and Kanehara, M. 2006. The question of prehistoric plant husbandry during the Jomon period in Japan. *World Archaeology*, 38(2):259–73.

Maxham, M. D. 2000. Rural communities in the Black Warrior valley, Alabama: The role of commoners in the creation of the Moundville I landscape. *American Antiquity*, 65(2):337–54.

Mayor, A., Huysecom, E., Gallay, A., Rasse, M., and Ballouche, A. 2005. Population dynamics and paleoclimate over the past 3000 years in the Dogon Country, Mali. *Journal of Anthropological Archaeology*, 24(1):25–61.

McClure, S. B. 2007. Gender, technology, and evolution: Cultural inheritance theory and prehistoric potters in Valencia, Spain. *American Antiquity*, 72(3):485–508.

McCoy, M. D., and Graves, M. W. 2010. The role of agricultural innovation on Pacific Islands: A case study from Hawai'i Island. *World Archaeology*, 42(1):90–107.

McGuire, K. R., and Hildebrandt, W. R. 2005. Re-thinking Great Basin foragers: Prestige hunting and costly signaling during the Middle Archaic period. *American Antiquity*, 70(4):695–712.

McNabb, J., Binyon, F., and Hazelwood, L. 2004. The large cutting tools from the South African Acheulean and the question of social traditions. *Current Anthropology*, 45(5):653–77.

Miller, N. F., Zeder, M., A., and Arter, S. R. 2009. From food and fuel to farms and flocks. *Current Anthropology*, 50(6):915–24.

Mills, B. J. 2007. Performing the feast: Visual display and suprahousehold commensalism in the Puebloan Southwest. *American Antiquity*, 72(2):210–39.

Mizoguchi, K. 2009. Nodes and edges: A network approach to hierarchisation and state formation in Japan. *Journal of Anthropological Archaeology*, 28(1):14–26.

Monaghan, G. W., and Peebles, C. S. 2010. The construction, use, and abandonment of Angel site mound A: Tracing the history of a Middle Mississippian town through its earthworks. *American Antiquity*, 75(4):935–53.

Moncel, M.-H., Brugal, J.-P., Prucca, A., and Lhomme, G. 2008. Mixed occupation during the Middle Palaeolithic: Case study of a small pit-cave-site of Les Pêcheurs (Ardèche, southeastern France). *Journal of Anthropological Archaeology*, 27(3):382–98.

Monnier, G. F. 2006. The Lower/Middle Paleolithic periodization in western Europe: An evaluation. *Current Anthropology*, 47(5):709–44.

Mora, R., and de la Torre, I. 2005. Percussion tools in Olduvai beds I and II (Tanzania): Implications for early human activities. *Journal of Anthropological Archaeology*, 24(2):179–92.

Morin, J. 2010. Ritual architecture in prehistoric complex hunter-gatherer communities: A potential example from Keatley Creek, on the Canadian Plateau. *American Antiquity*, 75(3):599–625.

Munro, N. D. 2004. Zooarchaeological measures of hunting pressure and occupation intensity in the Natufian: Implications for agricultural origins. *Current Anthropology*, 45(S4):S5–S34.

Munson, J. L., and Macri, M. J. 2009. Sociopolitical network interactions: A case study of the Classic Maya. *Journal of Anthropological Archaeology*, 28(4):424–38.

Nami, H. G. 2007. Research in the middle Negro River basin (Uruguay) and the Paleoindian occupation of the Southern Cone. *Current Anthropology*, 48(1):164–76.

Nelson, M. C., and Hegmon, M. 2001. Abandonment is not as it seems: An approach to the relationship between site and regional abandonment. *American Antiquity*, 66(2):213–35.

Nelson, M. C., Hegmon, M., Kulow, S., and Schollmeyer, K. G. 2006. Archaeological and ecological perspectives on reorganization: A case study from the Mimbres region of the U.S. Southwest. *American Antiquity*, 71(3):403–32.

Neme, G., and Gil, A. 2009. Human occupation and increasing mid-Holocene aridity: Southern Andean perspectives. *Current Anthropology*, 50(1):149–63.

Neves, W. A., Gonzáalez-José, R., Hubbe, M., Renato, K., G. M. Araujo, A., and Oldemar, B. 2004. Early Holocene human skeletal remains from Cerca Grande, Lagoa Santa, central Brazil, and the origins of the first Americans. *World Archaeology*, 36(4):479–501.

Nocete, F., Lizcano, R., Peramo, A., and Gómez, E. 2010. Emergence, collapse and continuity of the first political system in the Guadalquivir basin from the fourth to the second millennium BC: The long-term sequence of Úbeda (Spain). *Journal of Anthropological Archaeology*, 29(2):219–37.

Nocete, F., Sáez, R., Nieto, J. M., Cruz-Auñón, R., Cabrero, R., Alex, E., and Bayona, M. R. 2005. Circulation of silicified oolitic limestone blades in South-Iberia (Spain and Portugal) during the third millennium B.C.: An expression of a core/periphery framework. *Journal of Anthropological Archaeology*, 24(1):62–81.

O'Gorman, J. A. 2010. Exploring the longhouse and community in tribal society. *American Antiquity*, 75(3):571–97.

O'Sullivan, A. 2003. Place, memory and identity among estuarine fishing communities: Interpreting the archaeology of early medieval fish weirs. *World Archaeology*, 35(3):449–68.

Ortman, S. G. 2000. Conceptual metaphor in the archaeological record: Methods and an example from the American Southwest. *American Antiquity*, 65(4):613–45.

Ortman, S. G., Varien, M. D., and Gripp, T. L. 2007. Empirical Bayesian methods for archaeological survey data: An application from the Mesa Verde region. *American Antiquity*, 72(2):241–72.

Ortmann, A. L. 2010. Placing the Poverty Point mounds in their temporal context. *American Antiquity*, 75(3):657–78.

Panja, S. 2003. Mobility strategies and site structure: A case study of Inamgaon. *Journal of Anthropological Archaeology*, 22(2):105–25.

Parkinson, W. A., and Duffy, P. R. 2007. Fortifications and enclosures in European prehistory: A cross-cultural perspective. *Journal of Archaeological Research*, 15(2):97–141.

Pauketat, T. R. 2003. Resettled farmers and the making of a Mississippian polity. *American Antiquity*, 68(1):39–66.

Pavao-Zuckerman, B. 2007. Deerskins and domesticates: Creek subsistence and economic strategies in the historic period. *American Antiquity*, 72(1):5–33.

Pearson, R. 2006. Jomon hot spot: Increasing sedentism in south-western Japan in the Incipient Jomon (14,000–9250 cal. BC) and Earliest Jomon (9250–5300 cal. BC) periods. *World Archaeology*, 38(2):239–58.

Peregrine, P. N. 2001. Matrilocality, corporate strategy, and the organization of production in the Chacoan world. *American Antiquity*, 66(1):36–46.

Perry, J. E., and Jazwa, C. S. 2010. Spatial and temporal variability in chert exploitation on Santa Cruz Island, California. *American Antiquity*, 75(1):177–98.

Perttula, T. K., and Rogers, R. 2007. The evolution of a Caddo community in northeastern texas: The Oak Hill village site (41RK214), Rusk County, Texas. *American Antiquity*, 72(1): 71–94.

Peterson, C. E., and Drennan, R. D. 2005. Communities, settlements, sites, and surveys: Regional-scale analysis of prehistoric human interaction. *American Antiquity*, 70(1):5–30.

Peterson, J. 2010. Domesticating gender: Neolithic patterns from the southern Levant. *Journal of Anthropological Archaeology*, 29(3):249–64.

Pinhasi, R., and Pluciennik, M. 2004. A regional biological approach to the spread of farming in Europe: Anatolia, the Levant, south-eastern Europe, and the Mediterranean. *Current Anthropology*, 45(S4):S59–S82.

Pitts, M. 2008. Globalizing the local in Roman Britain: An anthropological approach to social change. *Journal of Anthropological Archaeology*, 27(4):493–506.

Pool, C. A. 2006. Current research on the Gulf Coast of Mexico. *Journal of Archaeological Research*, 14(3):189–241.

Porcasi, J. F., and Fujita, H. 2000. The dolphin hunters: A specialized prehistoric maritime adaptation in the southern California Channel Islands and Baja California. *American Antiquity*, 65(3):543–66.

Porcasi, J. F., Jones, T. L., and Raab, L. M. 2000. Trans-Holocene marine mammal exploitation on San Clemente Island, California: A tragedy of the commons revisited. *Journal of Anthropological Archaeology*, 19(2):200–220.

Potter, J. M. 2000. Pots, parties, and politics: Communal feasting in the American Southwest. *American Antiquity*, 65(3):471–92.

Potter, J. M., and Chuipka, J. P. 2010. Perimortem mutilation of human remains in an early village in the American Southwest: A case for ethnic violence. *Journal of Anthropological Archaeology*, 29(4):507–23.

Prentiss, A. M., Cross, G., Foor, T. A., Hogan, M., Markle, D., and Clarke, D. S. 2008. Evolution of a late prehistoric winter village on the interior plateau of British Columbia: Geophysical investigations, radiocarbon dating, and spatial analysis of the Bridge River site. *American Antiquity*, 73(1):59–81.

Prentiss, W. C., Lenert, M., Foor, T. A., Goodale, N. B., and Schlegel, T. 2003. Calibrated radiocarbon dating at Keatley Creek: The chronology of occupation at a complex hunter-gatherer village. *American Antiquity*, 68(4):719–35.

Price, T. D., Burton, J. H., and Stoltman, J. B. 2007. Place of origin of prehistoric inhabitants of Aztalan, Jefferson Co., Wisconsin. *American Antiquity*, 72(3):524–38.

Ramenofsky, A. F., Neiman, F. D., and Pierce, C. D. 2009. Measuring time, population, and residential mobility from the surface at San Marcos Pueblo, north central New Mexico. *American Antiquity*, 74(3):505–30.

Ray, H. P. 2004. The apsidal shrine in early Hinduism: Origins, cultic affiliation, patronage. *World Archaeology*, 36(3):343–59.

Redmond, B. G., and Tankersley, K. B. 2005. Evidence of early Paleoindian bone modification and use at the Sheriden Cave site (33WY252), Wyandot County, Ohio. *American Antiquity*, 70(3):503–26.

Reitz, E. J. 2004. "Fishing down the food web": A case study from St. Augustine, Florida, USA. *American Antiquity*, 69(1):63–83.

Richards, M. P., Price, T. D., and Koch, E. 2003. Mesolithic and Neolithic subsistence in Denmark: New stable isotope data. *Current Anthropology*, 44(2):288–95.

Richerson, P. J., Boyd, R., and Bettinger, R. L. 2001. Was agriculture impossible during the Pleistocene but mandatory during the Holocene? A climate change hypothesis. *American Antiquity*, 66(3):387–411.

Rick, T. C., Erlandson, J. M., and Vellanoweth, R. L. 2001. Paleocoastal marine fishing on the Pacific coast of the Americas: Perspectives from Daisy Cave, California. *American Antiquity*, 66(4):595–613.

Riel-Salvatore, J., Popescu, G., and Barton, C. M. 2008. Standing at the gates of Europe: Human behavior and biogeography in the southern Carpathians during the late Pleistocene. *Journal of Anthropological Archaeology*, 27(4):399–417.

Robbins, L. H., Campbell, A. C., Murphy, M. L., Brook, G. A., Srivastava, P., and Badenhorst, S. 2005. The advent of herding in southern Africa: Early AMS dates on domestic livestock from the Kalahari Desert. *Current Anthropology*, 46(4):671–77.

Roberts, B. 2008. Creating traditions and shaping technologies: Understanding the earliest metal objects and metal production in western Europe. *World Archaeology*, 40(3):354–72.

Roberts, N., and Rosen, A. 2009. Diversity and complexity in early farming communities of Southwest Asia: New insights into the economic and environmental basis of Neolithic Çatalhöyük. *Current Anthropology*, 50(3):393–402.

Robin, C. 2003. New directions in Classic Maya household archaeology. *Journal of Archaeological Research*, 11(4):307–56.

———. 2006. Gender, farming, and long-term change: Maya historical and archaeological perspectives. *Current Anthropology*, 47(3):409–33.

Robinson, B. S., Ort, J. C., Eldridge, W. A., Burke, A. L., and Pelletier, B. G. 2009. Paleoindian aggregation and social context at Bull Brook. *American Antiquity*, 74(3):423–47.

Rose, F. 2008. Intra-community variation in diet during the adoption of a new staple crop in the eastern woodlands. *American Antiquity*, 73(3):413–39.

Rosen, S. A., Savinetsky, A. B., Plakht, Y., Kisseleva, N. K., Khassanov, B. F., Pereladov, A. M., and Haiman, M. 2005. Dung in the desert: Preliminary results of the Negev Holocene Ecology Project. *Current Anthropology*, 46(2):317–26.

Rosenswig, R. M. 2006. Sedentism and food production in early complex societies of the Soconusco, Mexico. *World Archaeology*, 38(2):330–55.

Rothman, M. S. 2004. Studying the development of complex society: Mesopotamia in the late fifth and fourth millennia BC. *Journal of Archaeological Research*, 12(1):75–119.

Russell, N., Martin, L., and Buitenhuis, H. 2005. Cattle domestication at Çatalhöyük revisited. *Current Anthropology*, 46(5):S101–S108.

Sakaguchi, T. 2007. Refuse patterning and behavioral analysis in a pinniped hunting camp in the Late Jomon period: A case study in layer V at the Hamanaka 2 site, Rebun Island, Hokkaido, Japan. *Journal of Anthropological Archaeology*, 26(1):28–46.

———. 2009. Storage adaptations among hunter-gatherers: A quantitative approach to the Jomon period. *Journal of Anthropological Archaeology*, 28(3):290–303.

Sanhueza, L., and Falabella, F. 2010. Analysis of stable isotopes: From the Archaic to the horticultural communities in central Chile. *Current Anthropology*, 51(1):127–36.

Sara-Lafosse, R. V.-C. 2007. Construction, labor organization, and feasting during the Late Archaic period in the central Andes. *Journal of Anthropological Archaeology*, 26(2):150–71.

Sassaman, K. E. 2004. Complex hunter-gatherers in evolution and history: A North American perspective. *Journal of Archaeological Research*, 12(3):227–80.

Sassaman, K. E., Blessing, M. E., and Randall, A. R. 2006. Stallings Island revisited: New evidence for occupational history, community pattern, and subsistence technology. *American Antiquity*, 71(3):539–65.

Saunders, J. W., Mandel, R. D., Sampson, C. G., Allen, C. M., Allen, E. T., Bush, D. A., Feathers, J. K., et al. 2005. Watson Brake, a Middle Archaic mound complex in northeast Louisiana. *American Antiquity*, 70(4):631–68.

Savage, S. H. 2001. Some recent trends in the archaeology of Predynastic Egypt. *Journal of Archaeological Research*, 9(2):101–55.

Scarborough, V. L., and Burnside, W. R. 2010. Complexity and sustainability: Perspectives from the ancient Maya and modern Balinese. *American Antiquity*, 75(2):327–63.

Schachner, G. 2001. Ritual control and transformation in middle-range societies: An example from the American Southwest. *Journal of Anthropological Archaeology*, 20(2):168–94.

———. 2010. Corporate group formation and differentiation in early Puebloan villages of the American Southwest. *American Antiquity*, 75(3):473–96.

Schepartz, L. A., Miller-Antonio, S., and Bakken, D. A. 2000. Upland resources and the early Palaeolithic occupation of southern China, Vietnam, Laos, Thailand and Burma. *World Archaeology*, 32(1):1–13.

Schollmeyer, K. G., and Turner, C. G. 2004. Dental caries, prehistoric diet, and the pithouse-to-pueblo transition in southwestern Colorado. *American Antiquity*, 69(3):569–82.

Schroeder, S. 2004. Current research on late precontact societies of the midcontinental United States. *Journal of Archaeological Research*, 12(4):311–72.

Schulting, R. J., and Richards, M. P. 2001. Dating women and becoming farmers: New palaeodietary and AMS dating evidence from the Breton Mesolithic cemeteries of Téviec and Hoëdic. *Journal of Anthropological Archaeology*, 20(3):314–44.

Sealy, J. 2006. Diet, mobility, and settlement pattern among Holocene hunter-gatherers in southernmost Africa. *Current Anthropology*, 47(4):569–95.

Sealy, J., and Pfeiffer, S. 2000. Diet, body size, and landscape use among Holocene people in the southern Cape, South Africa. *Current Anthropology*, 41(4):642–55.

Shapland, A. 2010. The Minoan lion: Presence and absence on Bronze Age Crete. *World Archaeology*, 42(2):273–89.

Shelach, G. 2006. Economic adaptation, community structure, and sharing strategies of households at early sedentary communities in northeast China. *Journal of Anthropological Archaeology*, 25(3):318–45.

Sherwood, S. C., Boyce, N. D., Asa, R. R., and Scott, C. M. 2004. Chronology and stratigraphy at Dust Cave, Alabama. *American Antiquity*, 69(3):533–54.

Shimada, I., Shinoda, K., Farnum, J., Corruccini, R., and Watanabe, H. 2004. An integrated analysis of pre-Hispanic mortuary practices: A Middle Sican case study. *Current Anthropology*, 45(3):369–402.

Shoocongdej, R. 2000. Forager mobility organization in seasonal tropical environments of western Thailand. *World Archaeology*, 32(1):14–40.

Singh, U. 2004. Cults and shrines in early historical Mathura (c. 200 BC–AD 200). *World Archaeology*, 36(3):378–98.

Small, D. B. 2009. The dual-processual model in ancient Greece: Applying a post-neoevolutionary model to a data-rich environment. *Journal of Anthropological Archaeology*, 28(2):205–21.

Smith, A., and Munro, N. D. 2009. A holistic approach to examining ancient agriculture: A case study from the Bronze and Iron Age Near East. *Current Anthropology*, 50(6):925–36.

Smith, C. S. 2003. Hunter-gatherer mobility, storage, and houses in a marginal environment: An example from the mid-Holocene of Wyoming. *Journal of Anthropological Archaeology*, 22(2):162–89.

Smith, G. M. 2010. Footprints across the black rock: Temporal variability in prehistoric foraging territories and toolstone procurement strategies in the western Great Basin. *American Antiquity*, 75(4):865–85.

Smith, M. L. 2006. The archaeology of South Asian cities. *Journal of Archaeological Research*, 14(2):97–142.

Spencer, C. S., and Redmond, E. M. 2001. Multilevel selection and political evolution in the valley of Oaxaca, 500–100 B.C. *Journal of Anthropological Archaeology*, 20(2):195–229.

Srinivasan, S. 2004. Shiva as "cosmic dancer": On Pallava origins for the Nataraja bronze. *World Archaeology*, 36(3):432–50.

Stafford, C. R., Richards, R. L., and Anslinger, C. M. 2000. The bluegrass fauna and changes in middle Holocene hunter-gatherer foraging in the southern Midwest. *American Antiquity*, 65(2):317–36.

Stahl, A. B., Dores Cruz, M. D., Neff, H., Glascock, M. D., Speakman, R. J., Giles, B., and Smith, L. 2008. Ceramic production, consumption and exchange in the Banda area, Ghana: Insights from compositional analyses. *Journal of Anthropological Archaeology*, 27(3):363–81.

Stevenson, C. M., Ihab, A., and Steven, W. N. 2004. High precision measurement of obsidian hydration layers on artifacts from the Hopewell site using secondary ion mass spectrometry. *American Antiquity*, 69(3):555–67.

Steyn, M. 2003. A comparison between pre- and post-colonial health in the northern parts of South Africa, a preliminary study. *World Archaeology*, 35(2):276–88.

Stiger, M. 2006. A Folsom structure in the Colorado mountains. *American Antiquity*, 71(2):321–51.

Stiner, M. C. 2002. Carnivory, coevolution, and the geographic spread of the genus *Homo*. *Journal of Archaeological Research*, 10(1):1–63.

Stiner, M. C., Munro, N. D., and Surovell, T. A. 2000. The tortoise and the hare: Small-game use, the broad-spectrum revolution, and Paleolithic demography. *Current Anthropology*, 41(1):39–79.

Sulas, F., Madella, M., and French, C. 2009. State formation and water resources management in the Horn of Africa: The Aksumite Kingdom of the northern Ethiopian highlands. *World Archaeology*, 41(1):2–15.

Sutter, R. C., and Cortez, R. J. 2005. The nature of Moche human sacrifice: A bio-archaeological perspective. *Current Anthropology*, 46(4):521–49.

Szabó, K., Brumm, A., and Bellwood, P. 2007. Shell artefact production at 32,000–28,000 BP in island Southeast Asia: Thinking across media? *Current Anthropology*, 48(5):701–23.

Tafuri, M. A., Bentley, R. A., Manzi, G., and di Lernia, S. 2006. Mobility and kinship in the prehistoric Sahara: Strontium isotope analysis of Holocene human skeletons from the Acacus Mts. southwestern Libya). *Journal of Anthropological Archaeology*, 25(3):390–402.

Tankersley, K. B., Waters, M. R., and Stafford, T. W. 2009. Clovis and the American mastodon at Big Bone Lick, Kentucky. *American Antiquity*, 74(3):558–67.

Tartaron, T. F. 2008. Aegean prehistory as world archaeology: Recent trends in the archaeology of Bronze Age Greece. *Journal of Archaeological Research*, 16(2):83–161.

Tayles, N., Domett, K., and Nelsen, K. 2000. Agriculture and dental caries? The case of rice in prehistoric Southeast Asia. *World Archaeology*, 32(1):68–83.

Teyssandier, N. 2008. Revolution or evolution: The emergence of the Upper Paleolithic in Europe. *World Archaeology*, 40(4):493–519.

Theler, J. L., and Boszhardt, R. F. 2006. Collapse of crucial resources and culture change: A model for the Woodland to Oneota transformation in the upper Midwest. *American Antiquity*, 71(3):433–72.

Thompson, J. C., Sugiyama, N., and Morgan, G. S. 2008. Taphonomic analysis of the mammalian fauna from Sandia Cave, New Mexico, and the "Sandia Man" controversy. *American Antiquity*, 73(2):337–60.

Thompson, V. D. 2009. The Mississippian production of space through earthen pyramids and public buildings on the Georgia coast, USA. *World Archaeology*, 41(3):445–70.

Thompson, V. D., and Turck, J. A. 2009. Adaptive cycles of coastal hunter-gatherers. *American Antiquity*, 74(2):255–78.

Toll, H. W. 2001. Making and breaking pots in the Chaco world. *American Antiquity*, 66(1): 56–78.

Trubitt, M. B. D. 2000. Mound building and prestige goods exchange: Changing strategies in the Cahokia Chiefdom. *American Antiquity*, 65(4):669–90.

Truncer, J. 2004. Steatite vessel age and occurrence in temperate eastern North America. *American Antiquity*, 69(3):487–513.

Twiss, K. C. 2008. Transformations in an early agricultural society: Feasting in the southern Levantine Pre-Pottery Neolithic. *Journal of Anthropological Archaeology*, 27(4):418–42.

Tykot, R. H., and Staller, J. E. 2002. The importance of early maize agriculture in coastal Ecuador: New data from La Emerenciana. *Current Anthropology*, 43(4):666–77.

Ubelaker, D. H., and Owsley, D. W. 2003. Isotopic evidence for diet in the seventeenth-century colonial Chesapeake. *American Antiquity*, 68(1):129–39.

Ur, J. A. 2010. Cycles of civilization in northern Mesopotamia, 4400–2000 BC. *Journal of Archaeological Research*, 18(4):387–431.

Van de Noort, R. 2003. An ancient seascape: The social context of seafaring in the Early Bronze Age. *World Archaeology*, 35(3):404–15.

van der Merwe, N. J., Williamson, R. F., Pfeiffer, S., Thomas, S. C., and Allegretto, K. O. 2003. The Moatfield ossuary: Isotopic dietary analysis of an Iroquoian community, using dental tissue. *Journal of Anthropological Archaeology*, 22(3):245–61.

Van Dyke, R. M. 2004. Memory, meaning, and masonry: The Late Bonito Chacoan landscape. *American Antiquity*, 69(3):413–31.

Vanhaeren, M., and d'Errico, F. 2005. Grave goods from the Saint-Germain-la-Rivière burial: Evidence for social inequality in the Upper Palaeolithic. *Journal of Anthropological Archaeology*, 24(2):117–34.

Vanmontfort, B. 2008. Forager–farmer connections in an "unoccupied" land: First contact on the western edge of LBK territory. *Journal of Anthropological Archaeology*, 27(2):149–60.

Varela, H. H., and Cocilovo, J. A. 2000. Structure of the prehistoric population of San Pedro de Atacama. *Current Anthropology*, 41(1):125–32.

Varien, M. D., and Ortman, S. G. 2005. Accumulations research in the southwest United States: Middle-range theory for big-picture problems. *World Archaeology*, 37(1):132–55.

Varien, M. D., Ortman, S. G., Kohler, T. A., Glowacki, D. M., and Johnson, C. D. 2007. Historical ecology in the Mesa Verde region: Results from the Village Ecodynamics Project. *American Antiquity*, 72(2):273–99.

Vaughn, K. J. 2006. Craft production, exchange, and political power in the pre-Incaic Andes. *Journal of Archaeological Research*, 14(4):313–44.

Vega, B. M. 2009. Prehispanic warfare during the Early Horizon and Late Intermediate period in the Huaura valley, Peru. *Current Anthropology*, 50(2):255–66.

Vehik, S. C. 2002. Conflict, trade, and political development on the southern Plains. *American Antiquity*, 67(1):37–64.

Walde, D. 2006. Sedentism and pre-contact tribal organization on the northern Plains: Colonial imposition or indigenous development? *World Archaeology*, 38(2):291–310.

Walsh, K., Richer, S., and de Beaulieu, J. L. 2006. Attitudes to altitude: Changing meanings and perceptions within a "marginal" Alpine landscape; The integration of palaeoecological and archaeological data in a high-altitude landscape in the French Alps. *World Archaeology*, 38(3):436–54.

Warrick, G. 2003. European infectious disease and depopulation of the Wendat-Tionontate (Huron-Petun). *World Archaeology*, 35(2):258–75.

Washburn, D. K., Crowe, D. W., and Ahlstrom, R. V. N. 2010. A symmetry analysis of design structure: 1,000 years of continuity and change in Puebloan ceramic design. *American Antiquity*, 75(4):743–72.

Weber, A. W., and Bettinger, R. 2010. Middle Holocene hunter-gatherers of Cis-Baikal, Siberia: An overview for the new century. *Journal of Anthropological Archaeology*, 29(4):491–506.

Weber, A. W., Link, D. W., and Katzenberg, M. A. 2002. Hunter-gatherer culture change and continuity in the middle Holocene of the Cis-Baikal, Siberia. *Journal of Anthropological Archaeology*, 21(2):230–99.

Wells, P. S. 2005. Creating an imperial frontier: Archaeology of the formation of Rome's Danube borderland. *Journal of Archaeological Research*, 13(1):49–88.

Whalen, M. E., and Minnis, P. E. 2003. The local and the distant in the origin of Casas Grandes, Chihuahua, Mexico. *American Antiquity*, 68(2):314–32.

Wheeler, R. J., Miller, J. J., McGee, R. M., Ruhl, D., Swann, B., and Memory, M. 2003. Archaic period canoes from Newnans Lake, Florida. *American Antiquity*, 68(3):533–51.

White, C. D., Spence, M. W., Longstaffe, F. J., and Law, K. R. 2004. Demography and ethnic continuity in the Tlailotlacan enclave of Teotihuacan: The evidence from stable oxygen isotopes. *Journal of Anthropological Archaeology*, 23(4):385–403.

Whitridge, P. 2001. Zen fish: A consideration of the discordance between artifactual and zooarchaeological indicators of Thule Inuit fish use. *Journal of Anthropological Archaeology*, 20(1):3–72.

Wilkinson, T. J. 2000. Regional approaches to Mesopotamian archaeology: The contribution of archaeological surveys. *Journal of Archaeological Research*, 8(3):219–67.

Wills, W. H. 2001. Ritual and mound formation during the Bonito phase in Chaco Canyon. *American Antiquity*, 66(3):433–51.

Wilson, G. D. 2010. Community, identity, and social memory at Moundville. *American Antiquity*, 75(1):3–18.

Windes, T. C., and McKenna, P. J. 2001. Going against the grain: Wood production in Chacoan society. *American Antiquity*, 66(1):119–40.

Winterhalder, B., Kennett, D. J., Grote, M. N., and Bartruff, J. 2010. Ideal free settlement of California's northern Channel Islands. *Journal of Anthropological Archaeology*, 29(4): 469–90.

Wolverton, S. 2005. The effects of the hypsithermal on prehistoric foraging efficiency in Missouri. *American Antiquity*, 70(1):91–106.

Woodward, P., and Woodward, A. 2004. Dedicating the town: Urban foundation deposits in Roman Britain. *World Archaeology*, 36(1):68–86.

Yao, A. 2010. Recent developments in the archaeology of southwestern China. *Journal of Archaeological Research*, 18(3):203–39.

Yerkes, R. W. 2005. Bone chemistry, body parts, and growth marks: Evaluating Ohio Hopewell and Cahokia Mississippian seasonality, subsistence, ritual, and feasting. *American Antiquity*, 70(2):241–65.

Yuan, J., and Flad, R. 2005. New zooarchaeological evidence for changes in Shang dynasty animal sacrifice. *Journal of Anthropological Archaeology*, 24(3):252–70.

Zangrando, A. F. 2009. Is fishing intensification a direct route to hunter-gatherer complexity? A case study from the Beagle Channel region (Tierra del Fuego, southern South America). *World Archaeology*, 41(4):589–608.

Zeder, M. A. 2009. The Neolithic macro-(r)evolution: Macroevolutionary theory and the study of culture change. *Journal of Archaeological Research*, 17(1):1–63.

Zilhão, J. 2007. The emergence of ornaments and art: An archaeological perspective on the origins of "behavioral modernity." *Journal of Archaeological Research*, 15(1):1–54.

Bibliography

Ahrens, C. D. 1998. *Essentials of Meteorology: An Invitation to the Atmosphere.* Detroit, MI: Wadsworth.

Alroy, J. 2008. Dynamics of origination and extinction in the marine fossil record. *Proceedings of the National Academy of Sciences*, 105(Supplement 1):11536–42.

Alvarez, L. W., Alvarez, W., Asaro, F., and Michel, H. V. 1980. Extraterrestrial cause for the Cretaceous-Tertiary extinction. *Science*, 208(4448):1095–1108.

Ammerman, A. J., and Feldman, M. V. 1974. On the "making" of an assemblage of stone tools. *American Antiquity*, 39(4):610–16.

Andrefsky, W., Jr. 2005. *Lithics: Macroscopic Approaches to Analysis.* Cambridge: Cambridge University Press.

Arnold, P. J., III. 2008. No time like the present. In S. Holdaway and L. Wandsnider, editors, *Time in Archaeology: Time Perspectivism Revisited*, 13–30. Salt Lake City: University of Utah Press.

Ascher, R. 1961. Analogy in archaeological interpretation. *Southwestern Journal of Anthropology*, 17:317–25.

———. 1968. Time's arrow and the archaeology of a contemporary community. In K. C. Chang, editor, *Settlement Archaeology*, 43–52. Palo Alto, CA: National Press Books.

Bailey, G. N. 1981. Concepts, time-scales and explanations in economic prehistory. In A. Sheridan and G. N. Bailey, editors, *Economic Archaeology: Towards an Integration of Ecological and Social Approaches*, 99–117. BAR International Series, vol. 96. Oxford: BAR.

———. 1983. Concepts of time in Quaternary prehistory. *Annual Review of Anthropology*, 12:165–92.

———. 1987. Breaking the time barrier. *Archaeological Review from Cambridge*, 6:5–20.

———. 2007. Time perspectives, palimpsests and the archaeology of time. *Journal of Anthropological Archaeology*, 26(2):198–223.

———. 2008. Time perspectivism: Origins and consequences. In S. Holdaway and L. Wandsnider, editors, *Time in Archaeology: Time Perspectivism Revisited*, 13–30. Salt Lake City: University of Utah Press.

Bailey, G. N., and Jamie, W. 1997. The Klithi deposits: Sedimentology, stratigraphy and chronology. In G. Bailey, editor, *Klithi: Palaeolithic Settlement and Quaternary Landscapes in Northwest Greece*, 61–94. Cambridge: McDonald Institute for Archaeological Research.

Baker, A. 2011. Simplicity. In E. N. Zalta, editor, *The Stanford Encyclopedia of Philosophy*. Summer 2011 ed. Metaphysics Research Lab, Stanford University. http://plato.stanford.edu/archives/fall2013/entries/simplicity.

Barton, C. M., and Riel-Salvatore, J. 2014. The formation of lithic assemblages. *Journal of Archaeological Science*, 46:334–52.

Bar-Yosef, O., and Van Peer, P. 2009. The *chaîne opératoire* approach in Middle Paleolithic archaeology. *Current Anthropology*, 50(1):103–31.

Baumhoff, M. A., and Heizer, R. F. 1959. Some unexploited possibilities in ceramic analysis. *Southwestern Journal of Anthropology*, 15:308–16.

Behrensmeyer, A. K., and Hook, Robert, W. 1992. Paleoenvironmental contexts and taphonomic modes. In A. K. Behrensmeyer, J. D. Damuth, W. A. DiMichele, R. Potts, H.-D. Sues, and S. L. Wing, editors, *Terrestrial Ecosystems through Time: Evolutionary Paleoecology of Terrestrial Plants and Animals*, 15–136. Chicago: University of Chicago Press.

Behrensmeyer, A. K., and Chapman, R. E. 1993. Models and simulations of time-averaging in terrestrial vertebrate accumulation. In S. M. Kidwell and A. K. Behrensmeyer, editors, *Taphonomic Approaches to Time Resolution in Fossil Assemblages*, 125–49. Knoxville: Paleontological Society.

Behrensmeyer, A. K., Kidwell, S., and Gastaldo, R. A. 2000. Taphonomy and paleobiology. *Paleobiology*, 26(4):103–47.

Bentley, R. A., Maschnerand, H. D., and Chippindale, C., editors. 2008. *Handbook of Archaeological Theories*. New York: Rowman and Littlefield.

Benton, Michael, L. 2002. Cope's rule. In M. Pagel, editor, *Encyclopedia of Evolution*, 185–86. Oxford: Oxford University Press.

Binford, L. R. 1962. Archaeology as anthropology. *American Antiquity*, 28(2):217–25.

———. 1964. A consideration of archaeological research design. *American Antiquity*, 29(4):425–41.

———. 1965. Archaeological systematics and the study of culture process. *American Antiquity*, 31(2):203–10.

———. 1967. Smudge pits and hide smoking: The use of analogy in archaeological reasoning. *American Antiquity*, 32(1):1–12.

———. 1968a. Archaeological perspectives. In S. R. Binford and L. R. Binford, editors, *New Perspectives in Archaeology*, 5–32. Chicago: Aldine.

———. 1968b. Methodological considerations of the archaeological use of ethnographic data. In R. B. Lee and I. DeVore, editors, *Man the Hunter*, 268–73. Chicago: Aldine.

———. 1977. General introduction. In L. R. Binford, editor, *For Theory Building in Archaeology: Essays on Faunal Remains, Aquatic Resources, Spatial Analysis, and Systematic Modeling*, 1–10. New York: Academic Press.

———. 1981. Behavioral archaeology and the "Pompei Premise." *Journal of Anthropological Research*, 37(3):195–208.

———. 1982. Objectivity—explanation—archaeology 1982. In C. Renfrew, M. J. Rowland, and B. A. Segraves, editors, *Theory and Explanation in Archaeology: The Southampton Conference*, 125–38. New York: Academic Press.

———. 2001. Where do research problems come from? *American Antiquity*, 66(4):669–78.

Binford, L. R., and Binford, S. R., editors. 1968. *New Perspectives in Archaeology*. Chicago: Aldine.

Bintliff, J. 1991. The contribution of an *Annaliste*/structural history approach to archaeology. In J. Bintliff, editor, *The Annales School and Archaeology*, 1–33. Leicester: Leicester University Press,.

Boas, F. 1902. Some problems in North American archaeology. *American Journal of Archaeology*, 6:1–6.

Bobrowsky, P. T., and Ball, B. F. 1989. The theory and mechanics of ecological diversity in archaeology. In R. D. Leonard and G. T. Jones, editors, *Quantifying Diversity in Archaeology*, 4–12. Cambridge: Cambridge University Press.

Bocquet-Appel, J.-P. 2002. Paleoanthropological traces of a Neolithic demographic transition. *Current Anthropology*, 43(4):637–50.

———. 2009. La transition démographique agricole au Néolithique. In J.-P. Demoule, editor, *La révolution Néolithique dans le monde*, 301–17. Paris: CNRS.

———. 2011. When the world's population took off: The springboard of the Neolithic demographic transition. *Science*, 333(6042):560–61.

Bocquet-Appel, J.-P., and Naji, S. 2006. Testing the hypothesis of a worldwide Neolithic demographic transition: Corroboration from American cemeteries. *Current Anthropology*, 47(2):341–65.

Bohor, B. F. 1990. Shock-induced microdeformations in quartz and other mineralogical indications of an impact event at the Cretaceous-Tertiary boundary. *Tectonophysics*, 171(1–4):359–72.

Boyd, R., and Richerson, P. J. 1985. *Culture and the Evolutionary Process*. Chicago: University of Chicago Press.

Brain, J. P. 1969. Winterville: A case study of prehistoric culture contact in the lower Mississippi valley. PhD diss., Yale University, New Haven, CT.

Brantingham, P. J., Surovell, T. A., and Waguespack, N. M. 2007. Modeling post-depositional mixing of archaeological deposits. *Journal of Anthropological Archaeology*, 26:517–40.

Braudel, F. 1980. *On History*. Chicago: University of Chicago Press.

Bronk, R. C. 2009. Bayesian analysis of radiocarbon dates. *Radiocarbon*, 51(1):337–60.

Brown, J. H. 1995. *Macroecology*. Chicago: University of Chicago Press.

Brown, J. H., and Maurer, B. A. 1987. Evolution of species assemblages: Effects of energetic constraints and species dynamics on the diversification of the North American avifauna. *American Naturalist*, 130(1):111–17.

———. 1989. Macroecology: The division of food and space among species on continents. *Science*, 243(4895):1145–50.

Bryson, R. U., Bryson, R. A., and Ruter, A. 2006. A calibrated radiocarbon database of late Quaternary volcanic eruptions. *Earth Discussions*, 1:123–24.

Bush, A. M., Powell, M. G., Arnold, W. S., Bert, T. M., and Daley, G. M. 2002. Time-averaging, evolution, and morphologic variation. *Paleobiology*, 28(1):9–25.

Butzer, K. W. 1982. *Archaeology as Human Ecology: Method and Theory for a Contextual Approach*. Cambridge: Cambridge University Press.

Campbell, N. R. 1920. *Physics: The Elements*. Cambridge: Cambridge University Press.

———. 1921. *What Is Science?* London: Methuen and Co..

Cavalli-Sforza, L. L., and Feldman, M. V. 1981. *Cultural Transmission and Evolution: A Quantitative Approach*. Princeton, NJ: Princeton University Press.

Chalmers, A. F. 2013. *What Is This Thing Called Science?* 4th ed. Indianapolis, IN: Hackett.

Chamberlain, T. C. 1890. The method of multiple working hypotheses. *Science*, 15(366):92–96.

Chang, K. C. 1967. Major aspects of the interrelationship of archaeology and ethnology. *Current Anthropology*, 8(3):227–43.

Charlesworth, B. 1994. *Evolution in Age-Structured Populations*. Cambridge: University of Cambridge Press.

Chenet, A.-L., Courtillot, V., Fluteau, F., Gérard, M., Quidelleur, X., Khadri, S. F. R., Subbarao, K. V., and Thordarson, T. 2009. Determination of rapid Deccan eruptions across the Cretaceous-Tertiary boundary using paleomagnetic secular variation, 2: Constraints from analysis of eight new sections and synthesis for a 3500-m-thick composite section. *Journal of Geophysical Research: Solid Earth*, 114(B6): B06103.

Clark, J. G. D. 1951. Folk-culture and the study of European prehistory. In W. F. Grimes, editor, *Aspects of Archaeology*, 49–65. London: Edwards.

Clarke, D. L. 1968. *Analytical Archaeology*. London: Methuen.

Cleland, C. E. 2001. Historical science, experimental science, and the scientific method. *Geology*, 29(11):987–90.

———. 2002. Methodological and epistemic differences between historical science and experimental science. *Philosophy of Science*, 69(3):447–51.

———. 2011. Prediction and explanation in historical natural science. *British Journal for the Philosophy of Science*, 62(3):551–82.

Cobb, C. R. 2014. The once and future archaeology. *American Antiquity*, 79(4):589–95.

Collins, M. B. 1975. Sources of bias in processual data: An appraisal. In J. Mueller, editor, *Sampling in Archaeology*, 26–32. Tucson: University of Arizona Press.

Cooks, S. F. 1972a. Can pottery residues be used as an index to population? In *Contributions of the University of California Archeological Research Facility*, 17–39. Miscellaneous Papers on Archaeology, no. 14. Berkeley: Department of Anthropology, University of California.

———. 1972b. *Prehistoric Demography*. Reading, MA: Addison-Wesley.

Cowgill, G. L. 1970. *Some Sampling and Reliability Problems in Archaeology*. Paris: Colloques internationaux du CNRS.

David, N. 1972. On the life span of pottery, type frequencies, and archaeological inference. *American Antiquity*, 37(1):141–42.

David, N., and Kramer, C. 2001. *Ethnoarchaeology in Action*. Cambridge: Cambridge University Press.

Davies, B., Holdaway, S. J., and Fanning, P. C. 2016. Modelling the palimpsest: An exploratory agent-based model of surface archaeological deposit formation in a fluvial Australian landscape. *Holocene*, 26(3):450–63.

Davis, M., Hutt, P., and Muller, R. A. 1984. Extinction of species by periodic comet showers. *Nature*, 308:715–17.

Davis, S. D., editor. 1989. *The Hidden Falls Site, Baranof Island, Alaska*. Aurora Monograph Series, no. 5. Anchorage: Alaska Anthropological Association.

de Barros, P. L. F. 1982. The effects of variable site occupation span on the results of frequency seriation. *American Antiquity*, 47(2):291–315.

deBoer, W. R. 1983. The archaeological record as preserved death assemblage. In A. S. Keene and J. Moore, editors, *Archaeological Hammers and Theories*, 19–36. New York: Academic Press.

de Lange, J. 2008. Time averaging and the structure of archaeological records: A case study. In S. Holdaway and L. Wandsnider, editors, *Time in Archaeology: Time Perspectivism Revisited*, 149–60. Salt Lake City: University of Utah Press.

d'Errico, F., Banks, W. E., Vanhaeren, M., Laroulandie, V., and Langlais, M. 2011. PACEA georeferenced radiocarbon database. *PaleoAnthropology*, 2011, 1–12.

Diamond, J. 1997. *Guns, Germs, and Steel: The Fates of Human Societies*. New York: W. W. Norton.

Dibble, H. L., Holdaway, S. J., Lin, S. C., Braun, D. R., Douglass, M. J., Iovita, R., McPherron, S. P., Olszewski, D. I., and Sandgathe, D. 2016. Major fallacies surrounding stone artifacts and assemblages. *Journal of Archaeological Method and Theory*, 24(3):1–39.

Dillehay, Tom, D. 1989. *Monte Verde: A Late Pleistocene Settlement in Chile*. Washington, DC: Smithsonian Institution Press.

Dobzhansky, T. 1937. *Genetics and the Origin of Species*. New York: Columbia University Press.

Dunnell, R. C. 1971. *Systematics in Prehistory*. Caldwell, NJ: Blackburn Press.

———. 1982. Science, social science, and common sense: The agonizing dilemma of modern archaeology. *Journal of Anthropological Research*, 38:1–25.

———. 1984. Methodological issues in contemporary Americanist archaeology. *PSA: Proceedings of the Biennial Meeting of the Philosophy of Science Association*, 1984:717–44.

Dunning, T. 2012. *Natural Experiments in the Social Sciences: A Design-Based Approach*. Cambridge: Cambridge University Press.

Eighmy, J. L., and Labelle, J. M. 1996. Radiocarbon dating of twenty-seven Plains complexes and phases. *Plains Anthropologist*, 41(144):53–69.

Eldredge, N. 1989. *Macro-evolutionary Dynamics: Species, Niches, and Adaptive Peaks*. New York: McGraw-Hill.

Endler, J. 1986. *Natural Selection in the Wild*. Princeton, NJ: Princeton University Press.

Erwin, Douglas, H. 2000. Macroevolution is more than repeated rounds of microevolution. *Evolution and Development*, 2(2):78–84.

———. 2006. Dates and rates: Temporal resolution in the deep time stratigraphic record. *Annual Review of Earth and Planetary Sciences*, 2006(34):569–90.

Evershed, R. P. 2008. Organic residue analysis in archaeology: The archaeological biomarker revolution. *Archaeometry*, 50(6):895–924.

Ferring, C. R. 1986. Rates of fluvial sedimentation: Implications for archaeological variability. *Geoarchaeology*, 1(3):259–74.

Flessa, K. W. 1975. Area, continental drift and mammalian diversity. *Paleobiology*, 1(2):189–94.

Flessa, K. W., and Kowalewski, M. 1994. Shell survival and time-averaging in nearshore and shelf environments: Estimates from the radiocarbon literature. *Lethaia*, 27:153–65.

Fletcher, R. 1992. Time perspectivism, Annales, and the potential of archaeology. In A. B. Knapp, editor, *Archaeology, Annales, and Ethnohistory*, 35–50. Cambridge: Cambridge University Press.

Fogelin, L. 2007. Inference to the best explanation: A common and effective form of archaeological reasoning. *American Antiquity*, 72(4):603–25.

Foley, R. A. 1981. A model of regional archaeological structure. *Proceedings of the Prehistoric Society*, 47:1–17.

Foote, M. 1996. Evolutionary patterns in the fossil record. *Evolution*, 50(1):1–11.

———. 2007. Symmetric waxing and waning of marine invertebrate genera. *Paleobiology*, 33(4):517–29.

Forber, P., and Griffith, E. 2001. Historical reconstruction: Gaining epistemic access to the deep past. *Philosophy and Theory in Biology*, 3.

Forman, S. L. 2000. Luminescence geochronology. In J. S. Noller, J. M. Sowers, and W. R. Lettis, editors, *Quaternary Geochronology: Methods and Applications*, 157–76. AGU Reference Shelf 4. Washington, DC: American Geophysical Union.

Forster, M., and Sober, E. 1994. How to tell when simpler, more unified, or less ad hoc theories will provide more accurate predictions. *British Journal for the Philosophy of Science*, 45(1):1–35.

Fraassen, B. C. van. 1980. *The Scientific Image*. Oxford: Clarendon Press.

Frankel, D. 1988. Characterizing change in prehistoric sequences: A view from Australia. *Archaeology in Oceania*, 23(2):41–48.

Frederick, C. D., and Krahtopoulou, A. 2000. Deconstructing agricultural terraces: Examining the influence of construction method on stratigraphy, dating and archaeological visibility. In P. Halstead and C. D. Frederick, editors, *Landscape and Land Use in Postglacial Greece*, 79–93. Sheffield: Sheffield University Press.

Freeman, L. G. J. 1968. A theoretical framework for interpreting archaeological materials. In R. B. Lee and I. DeVore, editors, *Man the Hunter*, 262–67. Chicago: Aldine.

Fürsich, F. T., and Aberhan, M. 1990. Significance of time-averaging for palaeocommunity analysis. *Lethaia*, 23(2):143–52.

Gallivan, M. D. 2002. Measuring sedentariness and settlement population: accumulations research in the Middle Atlantic region. *American Antiquity*, 67(3):535–57.

Gamble, C. 1999. *The Palaeolithic Societies of Europe*. Cambridge: Cambridge University Press.

Gamble, C., and Porr, M. 2005. From empty spaces to lived lives: Exploring the individual in the Paleolithic. In C. Gamble and M. Porr, editors, *The Hominid Individual in Context: Archaeological Investigations of Lower and Middle Palaeolithic Landscapes, Locales and Artefacts*, 1–12. London: Routledge.

Gardner, T. W., Jorgensen, D. W., Shuman, C., and Lemieux, C. R. 1987. Geomorphic and tectonic process rates: Effects of measured time interval. *Geology*, 15(3):259–61.

Garvey, R. 2018. Current and potential roles of archaeology in the development of cultural evolutionary theory. *Philosophical Transactions of the Royal Society of London B: Biological Sciences*, 373(1473): 20170057.

Gaston, K. J., and Blackburn, T. M. 2000. *Pattern and Process in Macroecology*. Malden, MA: Blackwell Science.

Gee, H. 1999. *In Search of Deep Time: Beyond the Fossil Record to a New History of Life*. Ithaca, NY: Cornell University Press.

Gifford-Gonzalez, D. 1991. Bones are not enough: Analogues, knowledge, and interpretive strategies in zooarchaeology. *Journal of Anthropological Archaeology*, 10(3):215–54.

Gill, R. B., Mayewski, P. A., Nyberg, J., Haug, G. H., and Peterson, L. C. 2007. Drought and the Maya collapse. *Ancient Mesoamerica*, 18(2):283–302.

Gingerich, P. D. 1983. Rates of evolution: Effects of time and temporal scaling. *Science*, 222(4620):159–61.

Glymour, C. N. 1984. Explanation and realism. In J. Leplin, editor, *Scientific Realism*, 173–92. Berkeley: University of California Press.

Goethe, J. W. 1970. *Theory of Colors*. Cambridge, MA: MIT Press.

Gould, R. A., and Watson, P. J. 1982. A dialogue on the meaning and use of analogy in ethnoarchaeological reasoning. *Journal of Anthropological Archaeology*, 1(4):355–81.

Gould, S. J. 1965. Is uniformitarianism necessary? *American Journal of Science*, 263(3):223–28.

———. 1980. The promise of paleobiology as a nomothetic, evolutionary discipline. *Paleobiology*, 6(1):96–118.

Gould, S. J., and Eldredge, N. 1977. Punctuated equilibria: The tempo and mode of evolution reconsidered. *Paleobiology*, 3:115–51.

Graham, Russell, W. 1993. Processes of time-averaging in the terrestrial vertebrate record. In S. M. Kidwell and A. K. Behrensmeyer, editors, *Taphonomic Approaches to Time Resolution in Fossil Assemblages*, 102–24. Knoxville, TN: Paleontological Society.

Grant, P., and Grant, R. 1986. *Ecology and Evolution of Darwin's Finches*. Princeton, NJ: Princeton University Press.

Grayson, D. K. 1984. *Quantitative Zooarchaeology: Topics in the Analysis of Archaeological Faunas*. New York: Academic Press.

Grayson, D. K., and Delpech, F. 1998. Changing diet breadth in the early Upper Palaeolithic of southwestern France. *Journal of Archaeological Science*, 25:1119–29.

Greathouse, G. A., Fleer, B., and Wessel, C. J. 1954. Chemical and physical agents of deterioration. In G. A. Greathouse and C. J. Wessel, editors, *Deterioration of Materials: Causes and Preventive Techniques*, 71–174. New York: Reinhold.

Gurven, M., and Kaplan, H. 2007. Longevity among hunter-gatherers: A cross-cultural examination. *Population and Development Review*, 33(2):321–65.

Haldane, J. B. S. 1949. Suggestions as to quantitative measurements of rates of evolution. *Evolution*, 3(1):51–66.

Harris, M. 1994. Cultural materialism is alive and well and won't go away until something better comes along. In R. Borofsky, editor, *Assessing Cultural Anthropology*, 62–76. New York: McGraw-Hill.

Harrison-Buck, E. 2014. Anthropological archaeology in 2013: The search for truth(s). *American Anthropologist*, 116(2):338–51.

Hawkes, C. 1954. Archaeological theory and method: Some suggestions from the Old World. *American Anthropologist*, 56:155–68.

Hayden, B., and Villeneuve, S. 2011. A century of feasting studies. *Annual Review of Anthropology*, 40:433–49.

Hegmon, M. 2003. Setting theoretical egos aside: Issues and theory in North American archaeology. *American Antiquity*, 68(2):213–43.

Hempel, C. G. 1965. *Aspects of Scientific Explanation*. New York: Free Press.

Henrich, J. 2001. Cultural transmission and the diffusion of innovations: Adoption dynamics indicate that biased cultural transmission is the predominate force in behavioral change. *American Anthropologist*, 103(4):992–1013.

Henrich, J., Heine, S. J., and Norenzayan, A. 2010. The weirdest people in the world? *Behavioral and Brain Sciences*, 33:61–135.

Hergovich, A., Schott, R., and Burger, C. 2010. Biased evaluation of abstracts depending on topic and conclusion: Further evidence of a confirmation bias within scientific psychology. *Current Psychology*, 29(3):188–209.

Higgs, E. S. 1968. Archaeology—where now? *Mankind*, 6(12):617–20.

Hildebrand, A. R., Penfield, G. T., Kring, D. A., Pilkington, M., Camargo Z. A., Jacobsen, S. B., and Boynton, W. V. 1991. Chicxulub crater: A possible Cretaceous/Tertiary boundary impact crater on the Yucatán Peninsula, Mexico. *Geology*, 19(9):867–71.

Hodder, I. 2000. Agency and individuals in long-term processes. In M.-A. Dobres and J. Robb, editors, *Agency in Archaeology*, 21–33. London: Routledge.

———. 2001. Introduction: A review of contemporary theoretical debates in archaeology. In I. Hodder, editor, *Archaeological Theory Today*, 1–13. Cambridge: Cambridge University Press.

Holdaway, S. J. 2008. Time in archaeology: An introduction. In S. Holdaway and L. Wandsnider, editors, *Time in Archaeology: Time Perspectivism Revisited*, 1–12. Salt Lake City: University of Utah Press.

Holdaway, S. J., and Wandsnider, L. 2006. Temporal scales and archaeological landscapes from the eastern desert of Australia and intermontane North America. In G. Lock and B. L.

Molyneaux, editors, *Confronting Scale in Archaeology: Issues of Theory and Practice*, 183–202. New York: Springer.

———, editors. 2008. *Time in Archaeology: Time Perspectivism Revisited*. Salt Lake City: University of Utah Press.

Hosfield, R. T. 2001. The Lower Paleolithic of the Solent: "Site" formation and interpretive frameworks. In F. Wenban-Smith and R. T. Hosfield, editors, *Palaeolithic Archaeology of the Solent River*, 85–97. Lithic Studies Society Occasional Paper no. 7. London: Lithic Studies Society.

———. 2005. Individuals among palimpsest data: Fluvial landscapes in southern England. In C. Gamble and M. Porr, editors, *The Hominid Individual in Context: Archaeological Investigations of Lower and Middle Palaeolithic Landscapes, Locales and Artefacts*, 220–43. London: Routledge.

Howell, F. C. 1968. Discussions (part vi, 31a): Hunters in archaeological perspective. In R. B. Lee and I. DeVore, editors, *Man the Hunter*, 287–89. Chicago: Aldine.

Hunt, G. 2004. Phenotypic variation in fossil samples: Modeling the consequences of time-averaging. *Paleobiology*, 30(3):426–43.

Jablonski, D. 1999. The future of the fossil record. *Science*, 284(5423):2114–16.

———. 2008. Species selection: Theory and data. *Annual Review of Ecology and Systematics*, 39:501–24.

Jackson, J. B. C., and Erwin, D. H. 2006. What can we learn about ecology and evolution from the fossil record? *Trends in Ecology and Evolution*, 21:322–28.

Jeffares, B. 2008. Testing times: Regularities in the historical sciences. *Studies in History and Philosophy of Science, Part C: Studies in History and Philosophy of Biological and Biomedical Sciences*, 39(4):469–75.

———. 2010. Guessing the future of the past. *Biology and Philosophy*, 25(1):125–42.

Jones, G. T., Grayson, D. K., and Beck, C. 1983. Artifact class richness and sample size in archaeological surface assemblages. In R. C. Dunnell and D. K. Grayson, editors, *Lulu Linear Punctuated: Essays in Honor of George Irving Quimby*, 55–73. Anthropological Papers, no. 72. Ann Arbor: Museum of Anthropology, University of Michigan.

Kahn, J. G. 2013. Anthropological archaeology in 2012: Mobility, economy, and transformation. *American Anthropologist*, 115(2):248–61.

Kelley, J. H., and Hanen, M. P. 1988. *Archaeology and the Methodology of Science*. Albuquerque: University of New Mexico Press.

Kidwell, S. M., and Behrensmeyer, A. K. 1993. Taphonomic approaches to time resolution in fossil assemblages: Introduction. In S. M. Kidwell and A. K. Behrensmeyer, editors, *Taphonomic Approaches to Time Resolution in Fossil Assemblages*, 1–8. Knoxville, TN: Paleontological Society.

Kidwell, S. M., and Bosence, D. W. J. 1991. Taphonomy and time-averaging of marine shelly faunas. In P. A. Allison and D. E. Briggs, editors, *Taphonomy: Releasing the Data Locked in the Fossil Record*, 115–209. Topics in Geobiology 9. New York: Peplum Press.

Kidwell, S. M., and Flessa, K. W. 1996. The quality of the fossil record: Populations, species, and communities. *Annual Review of Earth and Planetary Sciences*, 24:433–64.

Kidwell, S. M., and Holland, M. S. 2002. The quality of the fossil record: Implications for evolutionary analyses. *Annual Review of Ecology and Systematics*, 33:561–88.

Kieseppä, I. A. 1997. Akaike information criterion, curve-fitting, and the philosophical problem of simplicity. *British Journal for the Philosophy of Science*, 48(1):21–48.

Kintigh, K. W. 1989. Sample size, significance, and measures of diversity. In R. D. Leonard and G. T. Jones, editors, *Quantifying Diversity in Archaeology*, 25–36. Cambridge: Cambridge University Press.

Kintigh, K. W., Altschul, J. H., Beaudry, M. C., Drennan, R. D., Kinzig, A. P., Kohler, T. A., Limp, W. F., et al. 2014. Grand challenges for archaeology. *Proceedings of the National Academy of Sciences*, 111(3):879–80.

Klayman, J., and Ha, Y. W. 1987. Confirmation, disconfirmation, and information in hypothesis testing. *Psychological Review*, 94(2):211–28.

Knapp, A. B. 1992. Archaeology and Annales: Time, space, and change. In A. B. Knapp, editor, *Archaeology, Annales, and Ethnohistory*, 1–21. Cambridge: Cambridge University Press.

Koehler, J. L. 1993. The influence of prior beliefs on scientific judgments of evidence quality. *Organizational Behavior and Human Decision Processes*, 56:28–55.

Kokko, H. 2007. *Modelling for Field Biologists and Other Interesting People*. Cambridge: Cambridge University Press.

Kowalewski, M. 1996. Time-averaging, overcompleteness, and the geological record. *Journal of Geology*, 104(3):317–26.

Kowalewski, M., and Bambach, R. K. 2003. The limits of paleontological resolution. In P. J. Harries, editor, *Approaches in High-Resolution Stratigraphic Paleontology*, 1–48. New York: Kluwer Academic/ Plenum.

Kowalewski, M., Goodfriend, G. A., and Flessa, K. W. 1998. High-resolution estimates of temporal mixing within shell beds: The evils and virtues of time-averaging. *Paleobiology*, 24(3):287–304.

Kroeber, A. L. 1909. The archaeology of California. In F. Boas, editor, *Putnam Anniversary Volume*, 1–42. New York: Stechert.

Kuhn, S. L., and Clark, A. E. 2015. Artifact densities and assemblage formation: Evidence from Tabun Cave. *Journal of Anthropological Archaeology*, 38(9):8–16.

Kuhn, T. S. 1962. *The Structure of Scientific Revolutions*. Chicago: University of Chicago Press.

Kukla, A. 1998. *Studies in Scientific Realism*. New York: Oxford University Press.

———. 2001. Theoricity, underdetermination, and the disregard for bizarre scientific hypotheses. *Philosophy of Science*, 68:21–35.

Laudan, L., and Leplin, J. 1991. Empirical equivalence and underdetermination. *Journal of Philosophy*, 88(9):449–72.

Lightfoot, K. G., and Jewett, R. A. 1984. The occupation duration of Duncan. In K. G., Lightfoot, editor, *The Duncan Project: A Study of the Occupation Duration and Settlement Pattern of an Early Mogollon Pithouse Village*, 47–82. Anthropological Field Studies, no. 6. Tempe: Office of Cultural Resource Management, Department of Anthropology, Arizona State University.

Lucas, G. 2005. *The Archaeology of Time*. London: Routledge.

———. 2012. *Understanding the Archaeological Record*. New York: Cambridge University Press.

Lyman, R. L. 1944. *Vertebrate Taphonomy*. Cambridge: Cambridge University Press.

———. 2003. The influence of time averaging and space averaging on the application of foraging theory in zooarchaeology. *Journal of Archaeological Science*, 30(5):595–610.

———. 2007. Archaeology's quest for a seat at the high table of anthropology. *Journal of Anthropological Archaeology*, 26(2):133–49.

Lyman, R. L., and O'Brien, M. J. 2001. The direct historical approach, analogical reasoning, and theory in Americanist archaeology. *Journal of Archaeological Method and Theory*, 8(4):303–41.

Lyman, R. L., O'Brien, M. J., and Dunnell, R. C. 1997. *The Rise and Fall of Culture History*. New York: Plenum Press.

Lyman, R. L., O'Brien, M. J., and Schiffer, M. B. 2005. Publishing archaeology in *Science* and *Scientific American*, 1940–2003. *American Antiquity*, 70(1):157–67.

Lyman, R. L., VanPool, T. L., and O'Brien, M. J. 2009. The diversity of North American projectile-point types, before and after the bow and arrow. *Journal of Anthropological Archaeology*, 28:1–13.

Lynch, T. F. 1990. Glacial age man in South America? A critical review. *American Antiquity*, 55(1):12–36.

MacCormac, Earl, R. 1976. *Metaphor and Myth in Science and Religion*. Durham, NC: Duke University Press.

Mace, R., and Pagel, M. 1995. A latitudinal gradient in the density of human languages in North America. *Proceedings of the Royal Society, Biological Sciences B*, 261(1360):117–21.

Macleod, N., Rawson, P. F., Forey, P. L., Banner, F. T., Boudagher-Fadel, M. K., Bown, P. R., Burnett, J. A., et al. 1997. The Cretaceous-Tertiary biotic transition. *Journal of the Geological Society*, 154(2):265–92.

Mahoney, M. J. 1977. Publication prejudices: An experimental study of confirmatory bias in the peer review system. *Cognitive Therapy and Research*, 1(2):161–75.

Mahoney, M. J., and Kimper, T. P. 1976. From ethics to logic: A survey of scientists. In M. J. Mahoney, editor, *Scientist as Subject*, 187–93. Cambridge, MA: Ballinger.

Malinksy-Buller, A., Hovers, E., and Marder, O. 2011. Making time: "Living floors," "palimpsets" and site formation processes—a perspective from the open-air Lower Paleolithic site of Revadim Quarry, Israel. *Journal of Anthropological Archaeology*, 30:89–101.

Manning, K., Timpson, A., Colledge, S., and Shennan, S. 2014. The chronology of culture: A comparative assessment of European Neolithic dating approaches. *Antiquity*, 88(342):1065–80.

Marean, C. W., and Spencer, L. M. 1991. Impact of carnivore ravaging on zooarchaeological measures of element abundance. *American Antiquity*, 56(4):645–58.

Marshall, C. R. 1990. Confidence intervals on stratigraphic ranges. *Paleobiology*, 16(1):1–10.

———. 1994. Confidence intervals on stratigraphic ranges: Partial relaxation of the assumption of randomly distributed fossil horizons. *Paleobiology*, 20(4):459–69.

———. 1997. Confidence intervals on stratigraphic ranges with nonrandom distributions of fossil horizons. *Paleobiology*, 23(2):165–73.

Martin, R. E. 1999. *Taphonomy: A Process Approach*. Cambridge: Cambridge University Press.

Mathew, S., and Perreault, C. 2015. Behavioural variation in 172 small-scale societies indicates that social learning is the main mode of human adaptation. *Proceedings of the Royal Society, Biological Sciences B*, 282:20150061.

McElreath, R. 2016. *Statistical Rethinking: A Bayesian Course with Examples in R and Stan*. Boca Raton, FL: CRC Press.

McKinney, M. L. 1991. *Completeness of the Fossil Record: An Overview*. London: Belhaven Press.

Meadow, R. H. 1981. Animal bones: Problems for the archaeologist together with some possible solutions. *Paléorient*, 6:65–77.

Medvedev, M. V., and Melott, A. L. 2007. Do extragalactic cosmic rays induce cycles in fossil diversity? *Astrophysical Journal*, 664:879–89.

Mellars, P. 2006. Why did modern human populations disperse from Africa ca. 60,000 years ago? A new model. *Proceedings of the National Academy of Sciences*, 103(25):9381–86.

Melott, A. L., and Bambach, R. K. 2014. Analysis of periodicity of extinction using the 2012 geological timescale. *Paleobiology*, 40(2):177–96.

Meltzer, D. J. 1985. North American archaeology and archaeologists, 1879–1934. *American Antiquity*, 50(2):249–60.

———. 2004. Modeling the initial colonization of the Americas: Issues of scale, demography, and landscape learning. In C. M. Barton, G. A. Clark, D. R. Yesner, and G. A. Pearsons, editors, *The Settlement of the American Continents*, 123–37. Tucson: University of Arizona Press.

———. 2005. The seventy-year itch: Controversies over human antiquity and their resolution. *Journal of Anthropological Research*, 61(4):433–68.

Meltzer, D. J., Leonard, R. D., and Stratton, S. K. 1992. The relationship between sample size and diversity in archaeological assemblages. *Journal of Archaeological Science*, 19(4):375–87.

Mills, B. J. 1989. Integrating functional analysis of vessels and sherds through models of ceramic assemblage formation. *World Archaeology*, 21:133–47.

Montanari, A., Hay, R. L., Alvarez, W., Asaro, F., Michel, H. V., Alvarez, L. W., and Smit, J. 1983. Spheroids at the Cretaceous-Tertiary boundary are altered impact droplets of basaltic composition. *Geology*, 11(11):668–71.

Morehart, C. 2015. Archaeologies of the past and in the present in 2014: Materialities of human history. Year in review: Archaeology. *American Anthropologist*, 117(2):329–44.

Morgan, L. E., and Renne, P. R. 2008. Diachronous dawn of Africa's Middle Stone Age: New 40ar/39ar ages from the Ethiopian Rift. *Geology*, 36(12):967–70.

Muller, A., Clarkson, C., and Shipton, C. 2017. Measuring behavioural and cognitive complexity in lithic technology throughout human evolution. *Journal of Anthropological Archaeology*, 48:166–80.

Murray, P. 1980. Discard location: The ethnographic data. *American Antiquity*, 45(3):490–502.

Murray, T. 1999. A return to "Pompeii Premise." In T. Murray, editor, *Time and Archaeology*, 8–27. London: Routledge.

Murray, T., and Walker, M. J. 1988. Like WHAT? A practical question of analogical inferences and archaeological meaningfulness. *Journal of Anthropological Archaeology*, 7(3):248–87.

Nakazawa, Y. 2016. The significance of obsidian hydration dating in assessing the integrity of Holocene midden, Hokkaido, northern Japan. *Quaternary International*, 397:474–83.

Nash, D. T., and Petraglia, M. D. 1984. Natural disturbance processes: A preliminary report on experiments in Jemez Canyon, New Mexico. *Haliksa'i: University of New Mexico, Contributions to Anthropology*, 3:129–47.

———. 1987. Natural formation processes and the archaeological record: Present problems and future requisites. In D. T. Nash and M. D. Petraglia, editors, *Natural Formation Processes and the Archaeological Record*, 186–204. BAR International Series 352. Oxford: BAR.

Nelson, M. C. 1991. The study of technological organization: Lithic and tool diversity. In M. B. Schiffer, editor, *Archaeological Method and Theory*, 3:57–100. Tucson: University of Arizona Press.

Nelson, N. C. 1909. Shellmounds of San Francisco Bay region. *Publications in American Archaeology and Ethnology*, 7:309–56.

Nettle, D. 1998. Explaining global patterns of language diversity. *Journal of Anthropological Archaeology*, 17(4):354–74.

Neugebauer, O. 1975. *A History of Ancient Mathematical Astronomy*. New York: Springer-Verlag.

Nickerson, R. S. 1998. Confirmation bias: A ubiquitous phenomenon in many guises. *Review of General Psychology*, 2(2):175–220.

O'Brien, M. J., and Lyman, R. L. 1999. *Seriation, Stratigraphy, and Index Fossils: The Backbone of Archaeological Dating*. New York: Kluwer Academic / Plenum.

———. 2002. The epistemological nature of archaeological units. *Anthropological Theory*, 2(1):37–56.

———. 2003. *Cladistics and Archaeology*. Salt Lake City: University of Utah Press.

O'Brien, M. J., Lyman, R. L., and Schiffer, M. B. 2005. *Archaeology as a Process: Processualism and Its Progeny*. Salt Lake City: University of Utah Press.

Olivier, L. 2011. *The Dark Abyss of Time: Archaeology and Memory*. New York: Altamira Press.

Olszewski, T. 1999. Taking advantage of time averaging. *Paleobiology*, 25(2):226–38.

Ormes, B. 1973. Archaeology and ethnography. In C. Renfrew, editor, *The Explanation of Culture Change: Models in Prehistory*, 481–92. Pittsburgh: University of Pittsburgh Press.

Oswald, M. E., and Grosjean, S. 2004. Confirmation bias. In R. Pohl, editor, *Cognitive Illusions*, 79–96. Hove: Psychological Press.

Parsons, J. R. 1974. The development of a prehistoric complex society: A regional perspective from the Valley of Mexico. *Journal of Field Archaeology*, 1(1–2):81–108.

Parsons, J. R., Kintigh, K. W., and Gregg, S. A. 1983. *Archaeological Settlement Pattern Data from the Chalco, Xochimilco, Ixtapalapa, Texcoco, and Zumpango*. Technical Report no. 14. Ann Arbor: University of Michigan Museum of Anthropology.

Peregrine, P. N. 2001. *Outline of Archaeological Traditions*. New Haven, CT: Human Relations Area Files.

Perreault, C. 2011. The impact of site sample size on the reconstruction of culture histories. *American Antiquity*, 76(3):547–72.

———. 2012. The pace of cultural evolution. *PLoS ONE*, 7(9):e45150.

———. 2018. Time-averaging slows down rates of change in the archaeological record. *Journal of Archaeological Method and Theory*, 25(3):953–64.

Perreault, C., Brantingham, P. J., Kuhn, S. L., Wurz, S., and Xing, G. 2013. Measuring the complexity of lithic technology. *Current Anthropology*, 54(S8):S397–S406.

Pierce, K. L., and Irving, F. 2000. Obsidian hydration dating of quaternary events. In J. S. Noller, J. M. Sowers, and W. R. Lettis, editors, *Quaternary Geochronology: Methods and Applications*, 223–40. AGU Reference Shelf 4. Washington, DC: American Geophysical Union.

Plog, F. T. 1974. *The Study of Prehistoric Change*. New York: Academic Press.

Plog, S. 1978. Social interaction and stylistic similarity: Reanalysis. In M. B. Schiffer, editor, *Advances in Archaeological Method and Theory*, 1:143–82. New York: Academic Press.

Plummer, M. 2013. Rjags: Bayesian graphical models using MCMC. In *R Package Version*, 3.

Porčić, M. 2015. Exploring the effects of assemblage accumulation on diversity and innovation rate estimates in neutral, conformist, and anti-conformist models of cultural transmission. *Journal of Archaeological Method and Theory*, 22(4):1071–92.

Premo, L. S. 2014. Cultural transmission and diversity in time-averaged assemblages. *Current Anthropology*, 55(1):105–14.

Prentiss, A. M., Kuijt, I., and Chatters, J. C., editors 2009. *Macroevolution in Human Prehistory*. New York: Springer.

Preucel, R. W., editor. 1991. *Processual and Postprocessual Archaeologies: Multiple Ways of Knowing the Past*. Center for Archaeological Investigations, Occasional Paper no. 10. Carbondale: Southern Illinois University Press.

Preucel, R. W., and Hodder, I., editors. 1996. *Contemporary Archaeology in Theory: A Reader*. Oxford: Blackwell.

Princehouse, P. M. 2003. Mutant Phoenix: Macroevolution in Twentieth-Century Debates over Synthesis and Punctuated Evolution. PhD diss., Harvard University, Cambridge, MA.

Psillos, S. 1999. *Scientific Realism: How Science Tracks the Truth*. New York: Routledge.

Pykles, B. 2008. A brief history of historical archaeology in the United States. *SAA Archaeological Record*, 8(3):32–34.

Ramenofsky, A. F. 1998. The illusion of time. In A. F. Ramenofsky and A. Steffen, editors, *Unit Issues in Archaeology*, 3–17. Salt Lake City: University of Utah Press.

Rapp, G. R., and Hill, C. L. 2006. *Geoarchaeology: The Earth-Science Approach to Archaeological Interpretation*. New Haven, CT: Yale University Press.

Raup, D. M. 1979. Biases in the fossil record of species and genera. *Bulletin of Carnegie Museum of Natural History*, 13:85–91.

Raup, D. M., and Sepkoski, J. J., Jr. 1982. Mass extinctions in the marine fossil record. *Science*, 215(4539):1501–3.

———. 1984. Periodicity of extinctions in the geologic past. *Proceedings of the National Academy of Sciences*, 81(3):801–5.

Reid, J. J., Schiffer, M. B., and Rathje, W. L. 1975. Behavioral archaeology: Four strategies. *American Antiquity*, 77:864–69.

Reimer, P., Bard, E., Bayliss, A., Beck, J., Blackwell, P., Ramsey, C. B., Buck, C., et al. 2013. IntCal13 and Marine13 radiocarbon age calibration curves 0–50,000 years cal BP. *Radiocarbon*, 55(4):1869–87.

Reitz, E. J., and Wing, E. S. 2008. *Zooarchaeology*. 2nd ed. New York: Cambridge University Press, New Yok.

Renfrew, C. 1973. *Before Civilization: The Radiocarbon Revolution and Prehistoric Europe*. New York: Alfred A. Knopf.

Renne, P. R. 2000. Kar and 40ar/39ar dating. In J. S. Noller, J. M. Sowers, and W. R. Lettis, editors, *Quaternary Geochronology: Methods and Applications*, 77–100. AGU Reference Shelf 4. Washington, DC: American Geophysical Union.

Resch, K., Ernst, E., and Garrow, J. 2000. A randomized controlled study of reviewer bias against an unconventional therapy. *Journal of the Royal Society of Medicine*, 93(4):164–67.

Rhode, D. 1988. Measurement of archaeological diversity and the sample-size effect. *American Antiquity*, 53(4):708–16.

Rhode, D., Brantingham, P. J., Perreault, C., and Madsen, D. B. 2014. Mind the gaps: Testing for hiatuses in regional radiocarbon date sequences. *Journal of Archaeological Science*, 52:567–77.

Richerson, P. J., and Boyd, R. 2005. *Not by Genes Alone: How Culture Transformed Human Evolution*. Chicago: University of Chicago Press.

Rick, J. W. 1976. Downslope movement and archaeological intrasite spatial analysis. *American Antiquity*, 41(2):133–44.

Rogers, A. R. 2000. On equifinality in faunal analysis. *American Antiquity*, 65(4):709–23.

Rogers, E. M. 1995. *Diffusion of Innovations*. 5th ed. New York: Free Press.

Rohde, R. A., and Muller, R. A. 2005. Cycles in fossil diversity. *Nature*, 434:208–10.

Rosenbaum, P. R. 2002. *Observational Studies*. 2nd ed. New York: Springer.

Ryan, B., and Gross, N. C. 1943. The diffusion of hybrid seed corn in two Iowa communities. *Rural Sociology*, 8(1):15–24.

Sadler, P. M. 1981. Sediment accumulation rates and the completeness of stratigraphic sections. *Journal of Geology*, 89(5):569–84.

Schick, K. D. 1987a. Experimentally-derived criteria for assessing hydrologic disturbance of archaeological sites. In D. T. Nash and M. D. Petraglia, editors, *Natural Formation Processes and the Archaeological Record*, 86–107. BAR International Series 352. Oxford: BAR.

———. 1987b. Modeling the formation of early Stone Age artifact concentrations. *Journal of Human Evolution*, 16(7):789–807.

Schiffer, M. B. 1972. Archaeological context and systemic context. *American Antiquity*, 37(2):156–65.

———. 1974. The effects of occupation span on site content. In *The Cache River Archaeological Project: An Experiment in Contract Archaeology*, 265–69. Arkansas Archaeological Survey, Research Series, 8. Fayetteville: Arkansas Archaeological Survey.

———. 1975. Archaeology as behavioral science. *American Anthropologist*, 77:836–48.

———. 1976. *Behavioral Archaeology*. New York: Academic Press.

———. 1987. *Formation Processes of the Archaeological Record*. New York: Academic Press.

———. 1988. The structure of archaeological theory. *American Antiquity*, 53(3):461–85.

Schiffer, M. B., and Wells, S. J. 1978. The design of archaeological surveys. *World Archaeology*, 10:1–28.

Schindel, D. E. 1980. Microstratigraphic sampling and the limits of paleontologic resolution. *Paleobiology*, 6(4):408–26.

———. 1982a. The gaps in the fossil record. *Nature*, 297(27):282–84.

———. 1982b. Resolution analysis: A new approach to the gaps in the fossil record. *Paleobiolgy*, 8(4):340–53.

Semaw, S. 2000. The world's oldest stone artefacts from Gona, Ethiopia: Their implications for understanding stone technology and patterns of human evolution between 2.6–1.5 million years ago. *Journal of Archaeological Science*, 27(12):1197–214.

Sepkoski, D. 2005. Stephen Jay Gould, Jack Sepkoski, and the "quantitative revolution" in American paleobiology. *Journal of the History of Biology*, 38:209–37.

———. 2012. *Rereading the Fossil Record*. Chicago: University of Chicago Press.

Sepkoski, J. J., Jr. 1982. A compendium of fossil marine families. *Milwaukee Public Museum Contributions in Biology and Geology*, 51:1–125.

Seymour, D. J. 2010. Contextual incongruities, statistical outliers, and anomalies: Targeting inconspicuous occupational events. *American Antiquity*, 75(1):158–76.

Shea, J. H. 1982. Twelve fallacies of uniformitarianism. *Geology*, 10(9):455–60.

Shennan, S. 1989. Archaeology as archaeology or as anthropology? Clarke's *Analytical Archaeology* and the Binfords' *New Perspectives in Archaeology* 21 years on. *Antiquity*, 63(241):831–35.

———. 2002. *Genes, Memes, and Human History: Darwinian Archaeology and Cultural Evolution*. New York: Thames and Hudson.

Short, M. B., D'Orsogna, M. R., Brantingham, P. J., and Tita, G. 2009. Measuring and modeling repeat and near-repeat burglary effects. *Journal of Quantitative Criminology*, 25(3):325–39.

Shott, M. J. 1989a. Diversity, organization, and behavior in the material record: Ethnographic and archaeological examples. *Current Anthropology*, 30(3):283–315.

———. 1989b. On tool-class use lives and the formation of archaeological assemblages. *American Antiquity*, 54(1):9–30.

———. 1996. Mortal pots: On use life and vessel size in the formation of ceramic assemblages. *American Antiquity*, 61(3):463–82.

———. 2004. Modeling use-life distributions in archaeology using New Guinea Wola ethnographic data. *American Antiquity*, 69(2):339–55.

———. 2006. Formation theory's past and future: Introduction to the volume. In *Formation Theory in Archaeology: Readings from American Antiquity*, 1–16. Washington, DC: SAA Press.

———. 2008. Lower Paleolithic industries, time, and the meaning of assemblage variation. In S. Holdaway and L. Wandsnider, editors, *Time in Archaeology: Time Perspectivism Revisited*, 46–60. Salt Lake City: University of Utah Press.

———. 2010. Size-dependence in assemblage measures: Essentialism, materialism, and "she" analysis in archaeology. *American Antiquity*, 75(4):886–906.

———. 2015. Theory in archaeology: Morphometric approaches to the study of fluted points. In N. Goodale and J. W. Andrefsky, editors, *Lithic Technological Systems and Evolutionary Theory*, 48–60. Cambridge: Cambridge University Press.

Signor, P. W., and Lipps, J. H. 1982. Sampling bias, gradual extinction patterns, and catastrophes in the fossil record. *Geological Society of America Special Papers*, 190:291–96.

Smit, J., and Hertogen, J. 1980. An extraterrestrial event at the Cretaceous-Tertiary boundary. *Nature*, 285(5762):198–200.

Smith, E. A. 2000. Three styles in the evolutionary analysis of human behaviour. In L. Cronk, N. Chagnon, and W. Irons, editors, *Adaptation and Human Behavior*, 25–60. New York: Aldine de Gruyter.

Smith, J. M. 1984. Paleontology at the high table. *Nature*, 309(31):401–2.

Smith, M. A. (1955) 1998. The limitations of inference in archaeology. *Archaeological Newsletter*, 6(1):3–7. Reprinted in L. E. Babits and H. Van Tilburg, editors, *Maritime Archaeology*, 167–74. Boston: Springer. Citations refer to the reprint edition.

Smith, M. E. 1992. Braudel's temporal rhythms and chronology theory in archaeology. In A. B. Knapp, editor, *Annales, Archaeology, and Ethnohistory*, 25–36. New York: Cambridge University Press.

Smith, M. L. 2010. *A Prehistory of Ordinary People*. Tucson: University of Arizona Press.

Smith, R. J., and Wood, B. 2017. The principles and practice of human evolution research: Are we asking questions that can be answered? *Comptes rendus palevol*, 16(5–6):670–79.

Sober, E. 1988. *Reconstructing the Past: Parsimony, Evolution, and Inference*. Cambridge, MA: MIT Press.

Stanley, Steven, M. 1973. An explanation for Cope's rule. *Evolution*, 27(1):1–26.

Stern, N. 1993. The structure of the Lower Pleistocene archaeological record. *Current Anthropology*, 34(3):201–25.

———. 1994. The implication of time-averaging for reconstructing the land-use patterns of early-tool-using hominids. *Journal of Human Evolution*, 27:89–105.

———. 2008. Time averaging and the structure of late Pleistocene archaeological deposits in southwest Tasmania. In S. Holdaway and L. Wandsnider, editors, *Time in Archaeology: Time Perspectivism Revisited*, 134–48. Salt Lake City: University of Utah Press.

Stevenson, C. M., Ladefoged, T. N., and Novak, S. W. 2013. Prehistoric settlement chronology on Rapa Nui, Chile: Obsidian hydration dating using infrared photoacoustic spectroscopy. *Journal of Archaeological Science*, 40(7):3021–30.

St. George, R. A., Snyder, T. E., Dykstra, W. W., and Henderson, L. S. 1954. Biological agents of deterioration. In *Deterioration of Materials: Causes and Preventive Techniques*, 175–233. New York: Reinhold.

Strauss, D., and Sadler, P. M. 1989. Classical confidence intervals and Bayesian probability estimates for ends of local taxon ranges. *Mathematical Geology*, 21:411–27.

Sullivan, A. P. I. 1992. Investigating the archaeological consequences of short-duration occupations. *American Antiquity*, 57(1):99–115.

———. 2008a. Ethnoarchaeological and archaeological perspectives on ceramic vessels and annual accumulation rates of sherds. *American Antiquity*, 73(1):121–35.

———. 2008b. Time perspectivism and the interpretive potential of palimpsests. In S. Holdaway and L. Wandsnider, editors, *Time in Archaeology: Time Perspectivism Revisited*, 31–45. Salt Lake City: University of Utah Press.

Surovell, T. A. 2009. *Toward a Behavioral Ecology of Lithic Technology: Cases from Paleoindian Archaeology*. Tucson: University of Arizona Press.

Surovell, T. A., and Brantingham, P. J. 2007. A note on the use of temporal frequency distributions in studies of prehistoric demography. *Journal of Archaeological Science*, 34(11):1868–77.

Surovell, T. A., Finley, J. B., Smith, G. M., Brantingham, P. J., and Kelly, R. 2009. Correcting temporal frequency distributions for taphonomic bias. *Journal of Archaeological Science,* 36:1715–24.

Surovell, T. A., Toohey, J. L., Myers, A. D., LaBelle, J. M., Ahern, J. C. M., and Reisig, B. 2017. The end of archaeological discovery. *American Antiquity,* 82(2):288–90.

Sutherland, W. J. 2003. Parallel extinction risk and global distribution of languages and species. *Nature,* 423(6937):276–79.

Svensson, A., Nielsen, S. W., Kipfstuhl, S., Johnsen, S. J., Steffensen, J. P., Bigler, M., Ruth, U., and Röthlisberger, R. 2005. Visual stratigraphy of the North Greenland Ice Core Project (North-grip) ice core during the last glacial period. *Journal of Geophysical Research: Atmospheres,* 110(D2):D02108.

Taylor, Walter, W. 1948. *A Study of Archaeology.* Memoirs of the American Anthropological Association no. 69. Menasha, WI: American Anthropological Association.

Tolstoy, P. 2008. Supplementary material for barkcloth, Polynesia, and cladistics: An update. *Journal of the Polynesian Society,* 117(1):15–57.

Trigger, Bruce, G. 1978. *Time and Traditions.* New York: Columbia University Press.

———. 1989. *A History of Archaeological Thought.* Cambridge: Cambridge University Press.

Trumbore, S. E. 2000. Radiocarbon geochronology. In J. S. Noller, J. M. Sowers, and W. R. Lettis, editors, *Quaternary Geochronology: Methods and Applications,* 41–60. AGU Reference Shelf 4. Washington, DC: American Geophysical Union.

Tucker, A. 2011. Historical science, over- and underdetermined: A study of Darwin's inference of origins. *British Journal for the Philosophy of Science,* 62(4):805–29.

Turchin, P. 2012. Dynamics of political instability in the United States, 1780–2010. *Journal of Peace Research,* 49(4):577–91.

Turner, D. 2005. Local underdetermination in historical science. *Philosophy of Science,* 72(1):209–30.

———. 2007. *Making Prehistory: Historical Science and the Scientific Realism Debate.* Studies in Philosophy and Biology. Cambridge: Cambridge University Press.

———. 2009. Beyond detective work: Empirical testing in paleontology. In D. Sepkoski and M. Ruse, editors, *The Paleobiological Revolution: Essays on the Growth of Modern Paleontology.* Chicago: University of Chicago Press.

———. 2011. *Paleontology: A Philosophical Introduction.* Cambridge: Cambridge University Press.

Turq, A., Roebroeks, W., Bourguignon, L., and Faivre, J.-P. 2013. The fragmented character of Middle Palaeolithic stone tool technology. *Journal of Human Evolution,* 65(5):641—655.

Underwood, B. D. 2002. Burden of proof. In J. Dressler, editor, *Encyclopedia of Crime and Justice,* 111–15. New York: Macmillan Reference USA.

Upham, S. 2004. The status and position of archaeology in American universities. *SAA Archaeological Record,* 4(2):6–8.

Utz, B., and Schyle, D. 2006. Near Eastern radiocarbon CONTEXT database, 2002–2006. http://www.context-database.de.

Valentine, J. W. 2009. The infusion of biology into paleontological research. In D. Sepkoski and M. Ruse, editors, *The Paleobiological Revolution: Essays on the Growth of Modern Paleontology,* 385–97. Chicago: University of Chicago Press.

Valentine, J. W., and Moores, E. M. 1970. Plate-tectonic regulation of faunal diversity and sea level: A model. *Nature,* 228(14):657–59.

Vaquero, M. 2008. The history of stones: Behavioural inferences and temporal resolution of an archaeological assemblage from the Middle Palaeolithic. *Journal of Archaeological Science,* 35(12):3178–85.

Varien, M. D., and Mills, B. J. 1997. Accumulations research: Problems and prospects for estimating site occupation span. *Journal of Archaeological Method and Theory,* 4(2):141–91.

Varien, M. D., and Ortman, S. G. 2005. Accumulations research in the southwest United States: Middle-range theory for big-picture problems. *World Archaeology,* 37(1):132–55.

Varien, M. D., and Potter, J. M. 1997. Unpacking the discard equation: Simulating the accumulation of artifacts in the archaeological record. *American Antiquity,* 62(2):194–213.

von Bertalanffy, L. 1940. Der Organismus als physikalisches System Betrachtet. *Naturwissenschaften,* 28:521–31.

———. 1949. Problems of organic growth. *Nature,* 163:156–58.

Waddington, C. H. 1977. *Tools for Thought.* New York: Basic Books.

Walker, K. R., and Bambach, R. K. 1971. The significance of fossil assemblages from fine-grained sediments: Time-averaged communities. *Geological Society of America Abstracts with Programs,* 3:783–84.

Wandsnider, L. 1987. Natural formation process experimentation and archaeological analysis. In D. T. Nash and M. D. Petraglia, editors, *Natural Formation Processes and the Archaeological Record,* 150–85. BAR International Series 352. Oxford: BAR.

———. 2008. Time-averaged deposits and multitemporal processes in the Wyoming Basin, intermontane North America. In S. Holdaway and L. Wandsnider, editors, *Time in Archaeology: Time Perspectivism Revisited,*61–93. Salt Lake City: University of Utah Press.

Waters, M. R. 1986a. *The Geoarchaeology of Whitewater Draw, Arizona.* Anthropological Papers of the University of Arizona, no. 45. Tucson: University of Arizona Press.

———. 1986b. The Sulphur Spring stage and its place in New World prehistory. *Quaternary Research,* 25:251–56.

———. 1992. *Principles of Geoarchaeology: A North American Perspective.* Tucson: University of Arizona Press.

Weisberg, J. S. 1981. *Meteorology: The Earth and Its Weather.* Boston: Houghton Mifflin.

Weisstein, E. W. 2015. Square line picking, from Mathworld—a wolfram web resource. http://mathworld.wolfram.com/SquareLinePicking.html.

West, F. H. 1993. Review of palaeoenvironment and site content at Monte Verde. *American Antiquity,* 58(1):166–67.

Willey, G. R., and Sabloff, J. A. 1993. *A History of American Archaeology.* 3rd ed. New York: Freeman.

Wilson, M. V. H. 1988. Paleoscene #9. Taphonomic processes: Information loss and information gain. *Geoscience Canada,* 15(2):131–48.

Wimsatt, W. C. 1987. False models as means to truer theories. In M. H. Nitcki and A. Hoffman, editors, *Neutral Models in Biology,* 23–55. New York: Oxford University Press.

Wood, R. W., and Johnson, D. L. 1978. A survey of disturbance processes in archaeological site formation. *Advances in Archaeological Method and Theory,* 1:315–81.

Wroe, S., and Field, J. 2006. A review of the evidence for a human role in the extinction of Australian megafauna and an alternative interpretation. *Quaternary Science Reviews,* 25(21):2692–2703.

Wylie, A. 1982. An analogy by any other name is just as analogical: A commentary on the Gould-Watson dialogue. *Journal of Anthropological Archaeology,* 1(4):382–401.

———. 1985. The reaction against analogy. In M. B. Schiffer, editor, *Advances in Archaeological Method and Theory*, 63–111. New York: Academic Press.

Yellen, John, E. 1977. *Archaeological Approaches to the Present: Models for Reconstructing the Past.* New York: Academic Press.

Zielinksi, G. A., Mayewski, P. A., Meeker, L. D., Whitlow, M. S., and Twickler, M. S. 1996. A 110,000-yr record of explosive volcanism from the GISP2 (Greenland) ice core. *Quaternary Research*, 45:109–18.

Index

9 780226 630960